Flood Risk Management and the American River Basin

An Evaluation

Committee on Flood Control Alternatives
in the American River Basin

Water Science and Technology Board

Commission on Geosciences, Environment, and Resources

National Research Council

NATIONAL ACADEMY PRESS
Washington, D.C. 1995

NOTICE: The project that is the subject of this report was approved by the Governing Board of the National Research Council, whose members are drawn from the councils of the National Academy of Sciences, the National Academy of Engineering, and the Institute of Medicine. The members of the committee responsible for the report were chosen for their special competences and with regard for appropriate balance.

This report has been reviewed by a group other than the authors according to procedures approved by a Report Review Committee consisting of members of the National Academy of Sciences, the National Academy of Engineering, and the Institute of Medicine.

Support for this project was provided by the United States Army Corps of Engineers under Contract Number DACW05-93-C-0087.

Library of Congress Catalog Card No. 95-70294
International Standard Book Number 0-309-05334-X

Additional copies of this report are available from:

National Academy Press
2101 Constitution Avenue, N.W.
Box 285
Washington, D.C. 20055
800-624-6242
202-334-3313 (in the Washington Metropolitan Area)
B-661

The cover shows a floodplain map depicting 100- and 400-year floodplains of the Sacramento and American rivers with the 1995 level of protection provided by Sacramento Area Flood Control Agency and the Sacramento District, U.S. Army Corps of Engineers.

The National Academy of Sciences is a private, nonprofit, self-perpetuating society of distinguished scholars engaged in scientific and engineering research, dedicated to the furtherance of science and technology and to their use for the general welfare. Upon the authority of the charter granted to it by the Congress in 1863, the Academy has a mandate that requires it to advise the federal government on scientific and technical matters. Dr. Bruce M. Alberts is president of the National Academy of Sciences.

The National Academy of Engineering was established in 1964, under the charter of the National Academy of Sciences, as a parallel organization of outstanding engineers. It is autonomous in its administration and in the selection of its members, sharing with the National Academy of Sciences the responsibility for advising the federal government. The National Academy of Engineering also sponsors engineering programs aimed at meeting national needs, encourages education and research, and recognizes the superior achievements of engineers. Dr. Harold Liebowitz is president of the National Academy of Engineering.

The Institute of Medicine was established in 1970 by the National Academy of Sciences to secure the services of eminent members of appropriate professions in the examination of policy matters pertaining to the health of the public. The Institute acts under the responsibility given to the National Academy of Sciences by its congressional charter to be an adviser to the federal government and, upon its own initiative, to identify issues of medical care, research, and education. Dr. Kenneth I. Shine is president of the Institute of Medicine.

The National Research Council was organized by the National Academy of Sciences in 1916 to associate the broad community of science and technology with the Academy's purposes of furthering knowledge and advising the federal government. Functioning in accordance with general policies determined by the Academy, the Council has become the principal operating agency of both the National Academy of Sciences and the National Academy of Engineering in providing services to the government, the public, and the scientific and engineering communities. The Council is administered jointly by both Academies and the Institute of Medicine. Dr. Bruce M. Alberts and Dr. Harold Liebowitz are chairman and vice chairman, respectively, of the National Research Council.

Preface

The headwaters of California's American River lie in the high Sierra Nevada. The river's three forks descend through a series of challenging rapids amid scenic canyons to merge just before flowing into Folsom Reservoir, about 23 miles upstream of Sacramento. Below Folsom Dam, the American River flows largely between levees to its convergence with the Sacramento River close to downtown Sacramento. Since its founding at the time of the Gold Rush in the 1850s, Sacramento has been battling to protect itself from floods, even as the city has continued to expand within the floodplain. The Sacramento Metropolitan Statistical Area reached a population of 1.4 million in 1990, an increase of 75 percent since 1970. Much of this population lives behind levees along the American River. Today, there are plans for a major new expansion in the Natomas Basin, a 55,000 acre expanse of low-lying, former marshland—now drained for agriculture—that lies within a 41-mile ring of levees across the American River from downtown Sacramento.

In February 1986, Sacramento had another brush with flood disaster as northern California was swept by vicious winter storms. Inflow from the upper American River watershed poured into Folsom Reservoir faster than it could be released through Folsom Dam's normal outlets. After the breaching of an upstream cofferdam released a surge of water into the reservoir, the dam operators opened five high-level spillways, gradually raising downstream flows to 130,000 cubic feet per second (cfs), well above the 115,000 cfs maximum release target for the levees along the lower American River. Extensive scouring occurred; Sacramento was spared a major disaster only by easing of storm conditions. Heavy rains in early 1995 reminded us again about the area's vulnerability.

Since the 1986 near-catastrophe, flood planners at the local, regional, state, and federal levels have struggled to develop an acceptable and feasible set of measures to improve Sacramento's level of safety from American River floods. In an attempt to identify and evaluate the various available alternatives, in 1991 the Sacramento District of the U.S. Army Corps of Engineers (USACE) prepared the American River Watershed Investigation (ARWI). The report examined a range of possible flood hazard reduction measures, but one in particular sparked controversy—possible construction of a "dry dam" upstream from Folsom. Planners had hoped the dry dam option would be a compromise acceptable to everyone.

The dry dam concept, however, proved unacceptable to some stakeholders, who assailed both the 1991 ARWI and its accompanying environmental impact statement on many technical grounds. Meanwhile, some proponents of power, water supply, and irrigation continued to press for a multipurpose dam as was originally proposed and authorized for the Auburn site, but which was never built for a combination of technical and political reasons. The result has been a virtual impasse regarding agreement on what flood control project to propose to Congress.

In 1992, Congress directed USACE to reevaluate the flood control options available for the American River basin (P.L. 102-396, Section 9159). Simultaneously, Congress directed the Secretary of the Army to solicit the views of the National Academy of Engineering with respect to certain technical and policy issues. Pursuant to that mandate, the Water Science and Technology Board of the National Research Council established the Committee on Flood Control Alternatives in the American River Basin, which began work in October 1993.

From the outset, this has been an unusual and challenging task for a National Research Council committee. The committee was originally called upon to review the 1991 ARWI, a report that was virtually moot by the time we came into existence. So we received many briefings and other informal input to enlarge and update our consideration of the American River flood dilemma. We paid particular attention to the new risk and uncertainty methodology now used by USACE to evaluate proposed projects (see Chapter 4). Our task was further complicated by the fact that we were asked to report on our findings before release of the upcoming, revised Supplemental Information Report and Environmental Impact Statement, meaning we had little written material upon which to base our analysis.

As this is being written in March 1995, much is happening in the lower American River Basin relevant to both our study and USACE's own reevaluation:

• The 1992 act that launched these studies also directed the Sacramento District of the USACE and the Sacramento Area Flood Control Agency (SAFCA) to strengthen the Natomas levees, provided that ". . . such construction does not encourage the development of deep floodplains (within the Natomas Basin)."

• The city and county of Sacramento have completed a lengthy planning process to guide new development in Natomas and it largely disregards flood hazards.

• The Bureau of Reclamation and the Sacramento District are developing a reoperation plan to make Folsom Dam more responsive to a developing flood situation.

• SAFCA has established a Lower American River Task Force involving diverse stakeholders to develop a consensus plan for the redesign of the levees and banks along the lower American River.

• SAFCA also is conducting research on the potential effects of occasional inundation on vegetation in the upper American River canyons.

• The U.S. Environmental Protection Agency and the U.S. Fish and Wildlife Service are developing new water release requirements to improve water quality and anadromous fish habitat in the Sacramento River delta, which may affect the operation of Folsom Dam.

• At the national level, a significant report on the 1993 Midwest floods *Sharing the Challenge: Floodplain Management into the 21st Century*, appeared during our study and we have considered its recommendations in this report where applicable.

Meanwhile in January and March 1995, California experienced widespread flooding, which stimulated "rethinking" about the siting of new development in hazardous areas (see *The New York Times*, January 15, 1995). Across the globe, winter floods in Holland forced precautionary evacuation of large areas behind dikes long thought to be safe, thus highlighting the costs and uncertainties of living behind flood barriers even when they do ultimately survive (see *The New York Times*, February 5, 1995).

Against this backdrop of ever-shifting political and scientific context, the committee sought to provide a useful and relevant report. We sought to address technical and policy issues both of immediate relevance to the American River basin and broader national significance. We were not charged, nor have we undertaken, to propose any particular solution for the lower American River flood problem. That is the responsibility of the political process. In particular, we take no position on whether or not Auburn dam should be built in any form. (We do strongly urge, however, that if a dry dam is built, it should have operable gates for both safety and environmental reasons.) The issue of Auburn dam has dominated the public debate on American River flood protection over the past decade to the possible detriment of giving fuller consideration to other approaches. We are pleased to note that certain recent initiatives, including but not limited to those listed above, are now in progress that do not depend on the resolution of the issue of whether Auburn dam should be built.

I would personally like to thank my colleagues and fellow committee members for their cooperation, hard work, mutual respect, and enthusiasm. A more distinguished yet congenial group of professionals can scarcely be imagined.

On behalf of the committee and the Water Science and Technology Board, I would like to express our appreciation to the fine officers and staff of the U. S. Army Corps of Engineers with whom we have interacted over the past 18 months. Our particular thanks are extended to Bob Childs, who served as the key liaison to the committee from the Sacramento District, plus Merritt Rice, Jaime Merino, Rick Johnson, and all the USACE staff who briefed us on issues and responded to our questions. We also appreciate the assistance we received from other liaisons to the committee, especially Tim Washburn, Sacramento Area Flood Control Agency; Ron Stork, Friends of the River; George Qualley, California Department of Water Resources; and Ray Barsh, California Reclamation Board.

Special thanks should go also to Mary Beth Morris for her calm and efficient logistical support. Finally, we thank WSTB senior staff officer Chris Elfring for her expertise and good judgment in guiding us through the many rapids and shoals of this project.

Rutherford H. Platt
University of Massachusetts, Amherst

Contents

Flood Risk Management and the American River Basin

Summary

There is no doubt that Sacramento, California, and the surrounding metropolitan area face a significant flood risk from both the Sacramento River and the American River, which converge at the city's doorstep. More than 400,000 people and $37 billion worth of damageable property are vulnerable to flooding in the Sacramento area, including most of the city's central business district and the State Capitol complex. Although there is consensus that action is needed to reduce the level of risk while allowing reasonable use of floodplains, agreement on the appropriate target level and the approach to achieve it has eluded national, state, regional, and local decisionmakers.

In 1991 the Sacramento District of the U.S. Army Corps of Engineers[1] completed a study, the *American River Watershed Investigation* (ARWI), that reviewed the American River's contribution to the area's flood hazard, considered a range of flood control measures, and recommended a preferred flood control strategy (USACE, Sacramento District, 1991). The effort met significant criticism, some of it highly technical and some of it political. As a result, Congress directed the Sacramento District to reevaluate its analysis and gather additional input. In response to this congressional directive, additional study and

---

[1]USACE (U.S. Army Corps of Engineers) as used herein refers to actions taken by the Washington D.C. headquarters of the Corps of Engineers (e.g., agency-wide policies, procedures, etc.) or comments by the headquarters on subordinate activities by subordinate elements such as the Sacramento District. Field activities, reports, work in progress, meetings, etc. by the Sacramento District are identified as the Sacramento District unless and until specifically acted on by USACE.

planning have been done by local, state, and federal interests. These efforts—which continue as this report goes to press—have yielded a more comprehensive picture of the flood risk and a broader array of possible flood risk reduction alternatives. The information available today is more comprehensive and more detailed than that available in 1991.

But the fundamental dilemma remains unresolved: how do we balance the potential benefits, impacts, costs, and trade-offs associated with the identified alternatives and select the best management plan for the basin and its residents? This final decision lies not in the realm of science and engineering, but in the arena of public decisionmaking. It requires participants to set aside differences and seek commonalities. It requires weighing competing values. In the end, it will require leadership from local governments, the state of California, the U.S. Army Corps of Engineers, and Congress, as well as a sincere effort by the region's interest groups to agree among themselves about how to respond to the flood hazard.

THE COMMITTEE'S CHARGE

At the same time that Congress asked for a reevaluation of the potential flood control alternatives available to the Sacramento area, the nature of some of the criticisms caused members of Congress to seek an outside body to review the technical soundness of the analyses and related policy questions. Congress directed the Secretary of the Army to ask the National Academy of Engineering to form a special committee, the Committee on Flood Control Alternatives in the American River Basin, to review the 1991 ARWI, with attention to the contingency assumptions, hydrologic methods, and other engineering analyses used to support the seven flood control options presented. Significantly, the committee was not asked to recommend a preferred alternative; instead, it was asked to evaluate the scientific and engineering knowledge base on which the selection of a final strategy will ultimately be based. The committee also was asked to take a step back from the often acrimonious debate that has surrounded the American River planning process and provide insights of value to other regions in the nation that face similar problems—other areas where cities have grown in flood-prone areas and now face significant flood risks. The massive Midwest floods of 1993 and significant regional floods in 1994 and 1995 are reminders of how serious this issue is for the nation.

The committee's charge contains an inherent dilemma. Because there was great controversy surrounding the 1991 ARWI, the committee was asked specifically to review that document. But the controversy surrounding that document was so great that Congress simultaneously asked the Sacramento District to revise it. As a result, while the committee was gathering information for its analysis, efforts to improve the 1991 report were being made by the Sacramento District, the Sacramento Area Flood Control Agency (SAFCA), the Reclamation

Board of the State of California, and the State Department of Water Resources, among others.

Thus in this report the committee comments on the data, analysis, and methodologies used in the 1991 ARWI where they are still germane. In addition, where possible, it reviews the new analyses and methodologies being used to reevaluate Sacramento's flood risks and assess alternative flood risk management strategies. This has proven to be a difficult task because the committee's study, the Sacramento District's ongoing efforts, and parallel work through SAFCA continue to move along in near synchrony. Also, at this point there is little written documentation of the new work.

The majority of the information concerning the ongoing work was received informally. The committee spoke at length with technical staff from both the federal and the state agencies and tried to understand what methodologies and data were being employed in the current analysis. The committee also heard from a variety of interest groups. In 1994, a new document, *Alternatives Report: American River Watershed, California* (USACE, Sacramento District, 1994a), reached the committee in time to be considered, but this interim report lacked detail. For a true reevaluation of the Sacramento District's technical analysis, Congress might wish to request a review of the upcoming *Draft Supplemental Information Report and Environmental Documentation*, expected to be available in the summer of 1995, because that document will update the 1991 ARWI in detail.

THREE PREMISES

As the committee conducted this review of the Sacramento District's planning for flood control in the American River basin, it became clear that the members shared certain premises (i.e., assumptions believed to be true on the basis of experience and expertise) that influenced their thinking. These premises are (1) the belief that alleviation of Sacramento's flood risk is critical, (2) the belief that decisionmaking in the American River basin should not stand in isolation, but should be seen in light of national policy that stresses the use of multiple strategies to respond to flood hazards, and (3) the belief that technical matters cannot be neatly separated from policy judgments. These three premises are introduced here to provide a context for understanding the scope of the committee's review and the nature of this report's conclusions and recommendations. These introductory ideas are followed by brief overviews of the chapters of the report.

Alleviation of Sacramento's Flood Risk is Critical

Actions to alleviate Sacramento's ongoing flood risk are urgently needed. The flood-prone development in Sacramento is intense and of high value. This

Sacramento, California, is a city that grew literally at the edge of the American River and it has been plagued by recurring floods as a result. More than 400,000 people and $37 billion worth of damageable property are vulnerable to flooding in the area, including most of the city's central business district and the State Capitol complex. (Robert Childs, U.S. Army Corps of Engineers.)

situation occurred in response to historical influences and cannot now be reversed. Nearly 10 years have elapsed since the city's existing flood defenses were clearly proven to be inadequate in the flood of February 1986. Although the careful analyses necessary to support decisionmaking take time, especially if the process is to allow adequate public participation, there comes a point where talk must turn into action. For a variety of reasons, the public decisionmaking process has been blocked from reaching consensus on a feasible course of action to provide the Sacramento area with a higher level of security. Paradoxically, efforts to enhance protection for the largely undeveloped floodplain of the Natomas Basin have progressed further in Congress and locally than proposals affecting developed areas, including downtown Sacramento and the State Capitol complex. Ultimately, California and the nation need to reexamine their approaches to public decisionmaking. Widespread involvement by stakeholders and careful consideration of all options is of course necessary. But delay per se can be counterproductive, costly, and potentially dangerous.

National Flood Policies Urge Multiple Adjustments to Hazard

Flood control in California cannot be treated in isolation but must be treated as a part of a complex system of water control and use that has evolved over a long period under the auspices of many government agencies and in response to significant pressures. As the committee approached the task of assessing flood risk along the lower American River, it was aware of recent laws and policy reviews that reflect a broadening of our nation's response to floods over the past quarter-century. For decades, the predominant response to flood risk was to build large flood control projects—dams, reservoirs, levees, diversion channels—to store and restrain floodwaters. The adoption of the National Flood Insurance Act of 1968 marked a watershed in national policy on flood hazards because it established nonstructural measures—flood insurance, floodplain management, and selective acquisition—as mainstays of national flood policy. Additional nonstructural measures in widespread use today include flood forecasting, evacuation planning, public education, and floodproofing of individual commercial and residential structures located in floodplains.

More recently, the 1994 Unified National Program for Floodplain Management (FEMA, 1994) also called for a blend of strategies, from structural approaches to modify flooding to restoration of floodplains. A major evaluation of the Midwest floods of 1993 prepared by the Interagency Floodplain Management Review Committee (IFMRC, 1994) at the direction of the White House, which calls for "shared responsibility" among all levels of government and private interests in responding to flood hazards, also strongly supports the use of nonstructural measures such as relocation of structures out of floodplains and restoration of wetlands, where feasible.

As noted by the National Review Committee (1989):

> The present status of floodplain management does not encourage complacency. The record is mixed. There are encouraging trends, as with the number of communities having some form of floodplain regulations, but the rising toll of average annual flood losses has not been stopped or reversed. Some activities look more productive on paper than on the ground or in the real vulnerability of people. On balance, progress has been far short of what is desirable or possible, or what was envisaged at times when the current policies and activities were initiated.

Thus planners and decisionmakers should proceed with caution. No single technical or institutional "fix" is likely to be an adequate response to the lower American River flood hazard. Responsible federal, state, regional, and local officials must seek to identify a combination of policies and measures that will maximize flood reduction benefits while minimizing economic and environmental costs.

Technical Assessment Includes Policy Judgments

The charge to the committee was based on the premise that many of the criticisms of the 1991 ARWI were matters of technical dispute and that a technical judgment could be rendered about the merits of the critics' comments. Representatives of USACE, SAFCA, environmental groups, and Congress at different times emphasized that the committee should try to settle the technical debate, in order to let the political process make the public policy choices about the acceptable risk at Sacramento, including Natomas. However, the planning and design of a flood control program, although requiring complex modeling, engineering, and data manipulation, do not divide neatly into two parts, technical analysis and policy decisions. For example, even the most apparently technical computational concerns, such as what to assume about the likely coincidence of peak flows at the confluence of two rivers or about use of surcharge space, are based on a policy viewpoint about the acceptable risk of modeling error.

IDENTIFICATION AND EVALUATION OF ALTERNATIVES

As the committee conducted its review of the American River planning process, it noted that perhaps the most critical step in the development of a flood control strategy is the selection of alternatives for detailed analysis. The 1991 ARWI presented various alternative approaches to providing flood control for the American River basin, addressing level of protection provided, costs, expected benefits, and environmental impacts. The report was controversial, and some criticisms were based on the perceived failure of the Sacramento District to consider and evaluate a full range of effective alternatives, such as modification of the operation of Folsom Dam coupled with improvements in outlet capacity. In considering the issue of alternative flood control plans in both the 1991 ARWI and a more recent document, the 1994 Alternatives Report, the committee focused on four issues: (1) use of Folsom Reservoir, (2) the question of gates should a dam be built at the Auburn site, (3) the viability of the Deer Creek alternative, and (4) the adequacy of the nonstructural measures presented.

As detailed in Chapter 3, the committee concludes that the original 1991 ARWI was reasonably complete, especially as supplemented by the 1994 Alternatives Report. One concern that arose involved the operating policies employed at Folsom Dam. However, ongoing investigations are now exploring the more dynamic use of Folsom storage capacity. Another concern is the fact that the committee was unable to evaluate how Folsom reoperation was actually considered in the 1994 Alternatives Report, particularly what assumptions were used regarding the initial conditions. These concerns are expected to be addressed in upcoming documents; resolution of these questions should not slow the planning process.

The committee notes that Folsom Reservoir, despite its limitations, is the critical component in the flood-control system for Sacramento. Consequently, it is essential that it be operated as efficiently as possible, and thus the soon-to-be released Folsom Flood Management Plan is critical. It is also important that the operation plan for Folsom evolve as necessary in response to changes in the American River system.

Regarding possible construction of a dry dam at the Auburn site, the committee notes that, should a dam be built, operational gates are essential for dam safety and to provide flexibility in the dam's operation, allowing operators to coordinate with Folsom and other flood control facilities, and to minimize environmental impacts in the upper American River canyon by regulating drawdown.

ENVIRONMENTAL ISSUES

A key issue in the controversy of how to provide flood hazard reduction to the American River basin is how to minimize environmental impacts. Environmental issues were at the heart of many of the disagreements that resulted from the 1991 ARWI. Among the most contentious were the question of the adequacy of the report in assessing potential environmental damage and the uncertainty surrounding impacts of a detention dam in Auburn canyon.

Overall, the committee finds that from an environmental perspective the 1991 ARWI suffered from a lack of scientifically based descriptions of potential impacts and thus did not adequately support the decisionmaking process and help the public weigh the environmental impacts for the range of flood damage reduction alternatives presented. The report understated some environmental impacts, particularly in the upper canyon. The 1994 Alternatives Report, subsequent research, and a report from the Lower American River Task Force (SAFCA, 1994b) show significant improvement in understanding impacts, consideration of options, and minimization of impacts.

On the basis of the research to date, the major uncertainty is potential impacts on canyon slopes and vegetation from inundation behind an Auburn detention dam. If such a dam is to be seriously considered, the committee recommends the formation of a multidisciplinary research team to design and carry out a program to reduce this scientific uncertainty and recommend a gate design and operating strategy that could be followed to minimize environmental impacts.

RISK METHODOLOGY

USACE has adopted new risk and uncertainty analysis procedures that are an extension of the traditional paradigm for flood control project planning and community flood protection evaluation. The 1994 Alternatives Report indicates that the Sacramento District's analysis now considers varying degrees of uncertainty in the causes of flooding, such as inflow to Folsom Reservoir, regulated outflow-frequency relationships for Folsom Dam, river stages, and levee stability. The methodology computes the risk of flooding due to combinations of hydrologic events, hydrologic parameter uncertainty, uncertainty in reservoir operations, stage-discharge relations, and levee performance. USACE traditionally has included safety factors in its design of facilities and the specification of operating policies to address important hydraulic and operational uncertainties in flood control planning calculations; with its new risk and uncertainty analysis methodology, one can investigate the extent to which such safety factors are economically justified.

The committee concludes that the USACE risk and uncertainty procedures are an important initiative. The explicit recognition of modeling uncertainty should result in a better understanding of the uncertainty of flood risk and damage reduction estimates. This change in methodology is important to the American River planning process because the ongoing evaluation of flood control alternatives for the basin is one of the first applications of the methodology. It is almost certainly the most complex application yet attempted by USACE.

As discussed in Chapter 4, the new risk and uncertainty procedures, which directly include hydrologic uncertainties in the calculation of average flood risk and the average annual flood damages, tend to inflate those estimates. This tendency can yield benefit-cost calculations more favorable to project justification. The chapter suggests how risk, variability, and uncertainty in hydrologic, hydraulic, and economic processes should be conceptualized and how the calculations can be organized to avoid introducing such biases while still communicating residual risks and associated uncertainty.

The committee also questions the value of the system reliability index computed by the Sacramento District in its American River study. The 1994 Alternatives Report was found to be particularly confusing because no distinction was made between estimates of flood risk calculated with the traditional level of protection and those calculated with the new risk and uncertainty procedures. Such distinctions are important. USACE needs to develop a consistent scientific methodology and an effective vocabulary for communication of residual flood risks and uncertainties to technical and public audiences.

FLOOD RISK MANAGEMENT BEHIND LEVEES

History shows a close relationship between flood protection and development in flood-prone areas. From the mid-1930s to the late 1960s federally subsidized flood control projects such as levees and upstream storage were the prevalent form of national response, but nevertheless flood losses continued to rise because of continuing development on floodplains. The reasons are many and complex: floodplains can appear to be desirable building locations, and the hazards sometimes are not seen or are unavoidable. Sometimes, development actually is encouraged by federal protection. Once a levee is built to protect development in a floodplain, for instance, it opens the way for additional development, which in turn prompts demands for higher levels of protection. Such development can impose heavy burdens on society. Thus in this era of tightened budgets the question of who pays to support this "flood protection-development spiral" is becoming increasingly important.

One question in the American River basin is whether this flood protection-development spiral is the fate of the Natomas Basin. The Natomas Basin is a flat, marshy lowland of about 55,000 acres near Sacramento that lies entirely within the 100-year floodplains of the American River and the Sacramento River. Today the basin is surrounded by a 41-mile ring of levees and is devoted primarily to agriculture. The basin is now home to 35,000 people, but because of its prime location, it is projected to be a major growth area for new housing and commercial development. Although the existing levees lessen the flood risk to some degree, the Natomas Basin faces significant residual risk. The basin lies below the levels of the American and Sacramento rivers at flood stage and could fill like a bathtub in the event of a flood that breaches or overtops the levees.

According to plans prepared by local authorities, large portions of the basin are poised for development despite the unresolved and perhaps unresolvable issue of its flood hazard. Clearly, the Natomas Basin is well situated in terms of proximity to Sacramento, but it is poorly situated in terms of chronic flood risk. Improvements in the existing flood protection system, including the reoperation of Folsom Dam, levee expansion, and other improvements that are in progress or are foreseeable, can help reduce the risk, but significant residual risk will remain. Development within the Natomas Basin thus should be subject to prudent floodplain management requirements under federal, state, and local authority. In addition, the public should be informed of the residual flood risks despite the presence of the levee system.

WATER RESOURCES PLANNING AND DECISIONMAKING

The application of the USACE planning process to the search for acceptable flood control for the American River basin has illustrated the need for reforms in how such decisions are made, and the committee believes that the lessons of the American River can be transferred to other areas of the nation. Early decisions, such as Congress's direction to limit the project purpose to flood control, that were made ostensibly to lessen controversy and speed the process instead prolonged the debate because public interests desired a wider view. Indeed, many who commented on the 1991 ARWI were critical of its failure to consider any purpose other than flood control as a planning purpose and noted that this single-purpose approach precluded a more integrated approach to planning. In particular, the dispute over the proposed dry dam alternative has stalled the study process. Although progress is being made to mate environmental restoration concerns with improved levee stability and conveyance in the lower American River, in large part because of the work of the Lower American River Task Force facilitated by SAFCA, efforts to resolve disputes over alternatives and impacts in the upper American River have met with little success.

The current decisionmaking situation in the American River basin can be described as a diffusion of separate interests having access to numerous political and legal veto points, making it far easier to stop an activity than to move one forward. Despite some errors and problems with the planning process as implemented in the American River basin, the committee recognizes that USACE to date has been embroiled in larger California water controversies and at times technical complaints have been used as weapons in a policy dispute. The committee believes that in the American River context and similar situations USACE must make its work part of a shared planning process where the local sponsor, other agencies of the federal and state governments, and nongovernmental interests can cooperatively develop the data and models, understandings of risks and tradeoffs, formulation of alternatives, and consensus on the most appropriate alternative.

The American River situation is not unusual; USACE has frequently seen its recommendations challenged in recent years and thus needs to find ways to improve the planning process so it works more effectively in the future. Areas open to reform include (1) acceptable damages and the flood insurance program, (2) water project cost sharing, (3) communication of flood risk, (4) water project planning, and (5) water policy and management at the national level.

FINDINGS AND RECOMMENDATIONS

This committee's task was to evaluate the scientific and engineering knowledge on which the selection of a flood hazard reduction strategy for the lower American River will ultimately be based. The committee also endeavored to provide insights on public policies concerning flood hazard management that are of concern to the nation. In line with that dual charge, the committee offers findings and recommendations specific to the USACE planning process as applied to the American River basin, as well as some broader comments on the nature of flood risk assessment and its application nationwide.

The findings and recommendations presented in detail in Chapter 7 relate to (1) the identification and evaluation of alternatives, (2) environmental issues, (3) risk methodology, (4) flood risk management behind levees, (5) risk communication, and (6) water resources planning and decisionmaking. Some of the key issues are summarized here, but Chapter 7 provides a fuller treatment of the findings and recommendations.

• Overall, the committee finds that the 1991 American River Watershed Investigation, as supplemented by the 1994 Alternatives Report, was reasonably complete in its consideration of structural flood protection measures. Alternative assumptions could have been selected, but nothing of a degree that should call the overall results into question.

• The committee does not and can not judge whether construction of a dry dam at the Auburn site is the best approach to reducing Sacramento's flood risk. However, the committee strongly believes that if a dry dam is built it must contain operational gates to ensure management flexibility, protect public safety, and minimize environmental impacts. Environmental concerns are significant, and additional research is needed to understand the potential impacts of a dam on the canyon environment, particularly plant communities and slope stability. Such information could be used to help set operational guidelines so impacts of such a dam could be minimized. In addition, if a dam is built, the committee believes it should be used as a last line of defense to contain peak flows from extreme floods, thus reducing the frequency of impacts on the canyon.

• The new USACE risk and uncertainty procedures are an innovative and timely development. The explicit recognition of modeling uncertainty should result in a better understanding of the uncertainty of flood risk and damage reduction estimates. However, the committee is concerned about the specific ways in which uncertainty is currently represented and included in the calculation of average flood damages and the residual risk of flooding, and about USACE's ability to communicate information about flood risk and community vulnerability. USACE leadership is encouraged to convene an intra-agency workshop, including outside experts, to review the new risk and uncertainty procedures.

• The determination of the federal interest in construction of water management facilities has always been a complex process affected by many factors, such as societal goals, the nature of the problem to be addressed, and financial constraints. The rationale for federal interest in flood control in the American River basin should be reviewed, and Congress should explicitly address whether federal involvement is warranted on the basis of the presence of widespread national benefits from flood protection or a limited ability of the community to provide its own flood protection. If a federal interest is clear, project construction should be delayed until SAFCA, working with FEMA and private insurers, has a program to require that new development at Natomas and in the city purchase flood insurance at actuarially sound rates for the residual risk. Also, SAFCA should implement a flood hazard mitigation plan, to be part of the area's land use plans, that includes flood risk communication, flood warning systems, evacuation plans to reduce loss of life, highway and other infrastructure designs to facilitate evacuation, and floodproofing and elevation requirements.

The fundamental question in the American River planning process is how to reduce flood risk in the lower American River basin given a decisionmaking arena that includes significant scientific uncertainty and organized opposition to some of the possible risk reduction alternatives. This report discusses the uncertainties that confront floodplain managers and offers recommendations in many areas, including the need for additional research in some areas. But decision-makers, agency officials, and interest groups reading this report should not use calls for additional research as an excuse for not taking action. It is time to select and implement a flood risk reduction strategy for the American River basin. There are still areas where data and information are incomplete, particularly in our understanding of environmental impacts, but that should not forestall the decisionmaking process. The recommendations offered in this report are intended to improve the process, not delay it further.

THE ROLE OF SCIENCE IN THE DECISIONMAKING PROCESS

The issue decisionmakers face is how best to determine and then implement an acceptable flood risk management program for the American River basin. Beyond all the complexities and subtleties, the ultimate question is whether the flood damage reduction offered through a combination of measures not including a dam is acceptable, or whether a new upstream dam is judged to be necessary to reduce risk to an acceptable level. The committee cannot answer that question, in part because detailed technical analyses comparing the alternatives are still being developed (this information is expected in the Sacramento District's forthcoming *Supplemental Information Report*, scheduled to be available in the summer of 1995) and, importantly, because that judgment is beyond the committee's appropriate role. The public should be forewarned that even when the technical analyses are available, there will be no simple technical answer. Scientists and engineers can and should provide careful analyses and interpret the information so it is available to support decisionmakers, and they should be frank about uncertainties and risks. But the decision to be made should ultimately reflect more than technical factors; it should reflect economic considerations and value judgments pertaining to the appropriate use of natural resources, public monies, acceptable levels of risk, and willingness to accept constraints on land use. The final decision on these issues rests with the public and the political officials who represent them.

1

Introduction

THE AMERICAN RIVER BASIN

In western lore, it is said that "Whiskey is for drinking and water is for fighting." California, with its long, dry summers, has seen its water dammed, diverted, channeled, and fought over for years. Such conflicts over water can be expected to continue, and even increase, as more people (30 million state-wide and rising) and more uses (agriculture, residential, municipal, industrial, power, flood control, recreation, and environment) compete for a fixed, although renewable, supply.

Today most, if not all, of the water in California is highly regulated and controlled by a patchwork quilt of laws, regulations, institutions, and facilities. The states's water supply is now so manipulated and interconnected that any changes in management policies should take into account the broad physical and historical context of the affected region, and sometimes the whole state. This chapter provides a brief introduction to the American River basin for readers unfamiliar with the area and its need for flood protection. The first section provides background on the physical setting and historical context within in which any flood management policy in the area should be considered. The second section provides an overview of the planning and decisionmaking process used by the U.S. Army Corps of Engineers (USACE). The application of this process in the American River basin is described in more detail in Chapter 6.

Physical Setting

The American River Basin is located east of Sacramento in the northwestern Sierra Nevada (Figure 1.1). The watershed encompasses about 2,000 square miles. Elevations range from 10,400 feet in the high peaks on the Sierra crest to only 30 feet at Sacramento. A range of meteorological, topographic, and hydrologic conditions contribute to the basin's current flood problem.

The climatic regime of California is Mediterranean, with cool, wet winters and dry summers. At high elevations some modest summer precipitation occurs but does not generate regional flooding. The steep, west-facing slopes of the upper basin present an orographic barrier that extracts moisture from the prevailing maritime westerlies. Mean annual precipitation varies with elevation, forming a steep precipitation gradient from the Sacramento Valley up to the Sierra crest, from 18 to 70 in./yr, respectively (USACE, Sacramento District[1], 1991, Appendix K). Annual precipitation also varies greatly from year to year, and precipitation in the upper basin can be quite intense. For example, during the severe storms of 1986, rainfall intensities in the mountains reached as much as 0.75 in./hr, and many daily totals exceeded 10 inches (California Department of Water Resources, 1988).

Knowledge of the region's past climates remains qualitative and incomplete, introducing hydrologic uncertainty that cannot be quantitatively incorporated into a risk analysis. It is clear that climatic variability has been substantial. Dendroclimatologic data from 1560 to 1979 A.D. suggest that more recent years have been moist and that the 1930s represent the driest period of the entire record in the Sacramento basin (Earle, 1993). Prolonged departures from the mean are commonplace.

Factors affecting flood hydrology include geology, soils, vegetation, and artificial impoundments in the upper basin. Basin topography varies from extremely rugged in the mountains to very flat in the Sacramento Valley. Much of the upper basin above Folsom Dam drains into a network of deep ravines separated by high, steep-sided ridges. The drainage network can be divided into three primary branches: the North and Middle Forks, which meet near the town of Auburn, and the South Fork, which joins the American River at Folsom Reservoir (Figure 1.1). The steep, rocky canyons of the upper basin afford little natural storage of the intense rainfalls that may occur during the rainy season. Except in dense forest or where there is a deep snowpack, most precipitation is quickly

---

[1]USACE (U.S. Army Corps of Engineers) as used herein refers to actions taken by the Washington D.C. headquarters of the Corps of Engineers (e.g., Corps wide policies, procedures, etc.) or comments by the headquarters on subordinate activities by subordinate elements such as Sacramento District. Field activities, reports, work in progress, meetings, etc. by Sacramento District should be identified as the "District" or "Sacramento District" unless and until specifically acted upon officially by "USACE."

delivered to channels and conveyed downstream. The elevation of the snowpack, therefore, is critical to runoff response.

Vegetation in the American River basin is strongly related to topographic position and has much bearing on spatial characteristics of rainfall-runoff relationships. The upper third of the basin is dominated by glacially polished bedrock and thin vegetation ranging from alpine tundra to subalpine forest communities (Munz and Keck, 1973). Much of the basin is at moderate elevations, where gentle slopes are colonized by thick mixed coniferous forests. A grove of giant sequoia on the Middle Fork indicates the ample moisture available in the forest belt of the basin. In general, forested areas do not produce as much runoff as other surfaces. At lower elevations, vegetation thins out to grassland, chaparral, and woodland species in the foothills, and grassland savanna or riparian hardwoods in the Sacramento Valley.

Folsom Dam, the largest dam on the American River, has a low volume-to-runoff ratio, and given its current design and operations it is incapable of storing and then releasing the bulk of a major flood on the river. Several small privately owned reservoirs in the basin's upper tributaries are operated primarily for power generation. Five of these reservoirs account for about 90 percent of the total storage capacity above Folsom Dam and collectively control about 14 percent of the drainage area above Folsom (USACE, Sacramento District, 1991).

The lower basin is distinctly different from the upper basin. Below the town of Folsom, the American River emerges onto an alluvial plain with high, steeply dipping bluffs on the north side. Tributaries on the northern upland drain west-northwest to the Sacramento River. Downstream, below Rancho Cordova, the topography flattens out, and the American River ultimately joins the Sacramento River. Historically, the Sacramento area was marshy and prone to flooding in most years (John Work, 1833, as described in Dillinger, 1991; Lt. Derby, 1849, as described in Farquhar, 1932). Historical sedimentation by hydraulic gold mining altered the lower American River channel system from its natural state (Gilbert, 1917), but the area remained marshy, and a 1907 topographic map (California Debris Commission, 1907) represented the Natomas area as "Lake American."

Historical Context of the Flood Control Controversy

The Sacramento River has of course flooded since time immemorial. But the starting point of flood control in the Sacramento Valley was the decision of the City of Sacramento to remain in the floodplain after a major flood in 1850, rather than moving to higher ground (Table 1.1). As towns grew and prospered along the river, and larger landowners drained swamplands for agriculture, the prevention of flood damage became a dominant issue in the politics of the valley. Despite construction of significant flood control features (Figure 1.2) and a long series of studies, reports, and laws designed to reduce the area's risk, the Sacramento River has continued to experience devastating floods.

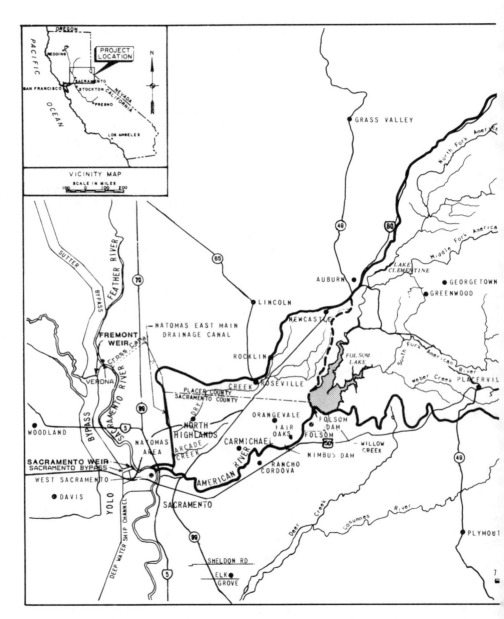

FIGURE 1.1 Main features of the American River Watershed. SOURCE: Sacramento District, USACE, 1991.

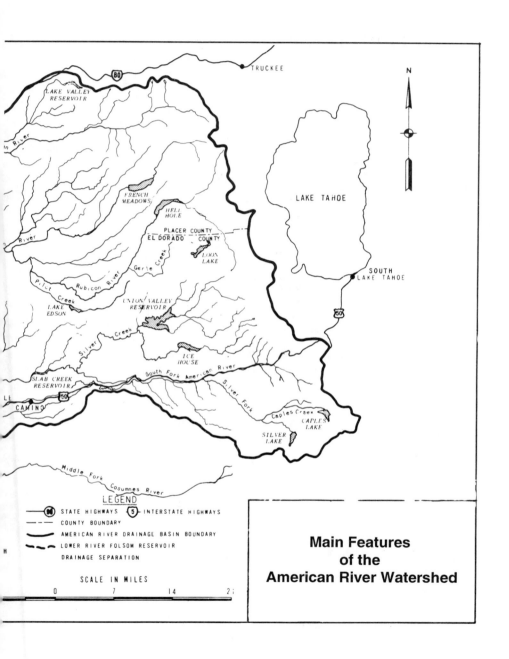

Main Features
of the
American River Watershed

TABLE 1.1 Chronology of Sacramento/American River Flood History

1848	Discovery of gold near Sacramento
1850	Major flood—Sacramento stays put, starts building levees
1861	Humphreys and Abbot report for the Mississippi River—"levees-only" policy
1862	Major flood—Sacramento begins to elevate streets and improve levees
1868	Green Act adopted—authorizes local levee districts
1881	Major flood—aggravated by hydraulic mining
1883	*Moulton v. Parks*—upholds suit against levee causing overflow onto adjacent land
1884	*Woodruff v. North Bloomington Gravel Mining Corp*—bans hydraulic mining
1891	Major flood—destroys hydraulic mining infrastructure in mountains
1893	California Debris Commission created by Congress; small-scale, licensed hydraulic mining resumes
1894	Debate between "levees only" and combination approach continues
1905	Sacramento Drainage District established
1907	Great flood exceeding 600,000 cfs peak flow—discredited "levees only" policy
1910	Jackson plan—levees, bypasses, channel widening
1911	California Legislature adopts Jackson plan
1917	Federal Flood Control Act adopts Jackson plan—50-50 cost sharing
1935	Central Valley Project authorized by Congress
1956	Folsom Dam completed
1962	Lower American River Parkway established
1965	Auburn dam authorized by Congress
1975	Auburn dam construction suspended due to Oroville earthquake
1986	Major flood—nearly overtops downstream levees at Sacramento
1991	American River Watershed Investigation Feasibility Report published
1992	P.L. 102-396 authorizes Natomas elements and mandates this study
1993	NRC Committee on Flood Control Alternatives in the American River Basin formed
1994	American River Alternatives Report published

Several long-standing issues continue to complicate present-day efforts to achieve safety from floods in the lower American River basin. These include:

- the scale of decisionmaking and the problem of externalities,
- competing strategies of flood management,
- intergovernmental cooperation and cost sharing, and
- scientific uncertainty.

The Scale of Decisionmaking and Externalities

The politics of flood control in the Sacramento Valley reflect a recurrent debate between the advocates of centralization and decentralization in decision-making. During the second half of the nineteenth century, California Republicans favored centralized management based on technical expertise, while Democrats favored a more laissez-faire approach. The latter prevailed when the state legislature adopted the Green Act in 1868, which authorized the creation of local

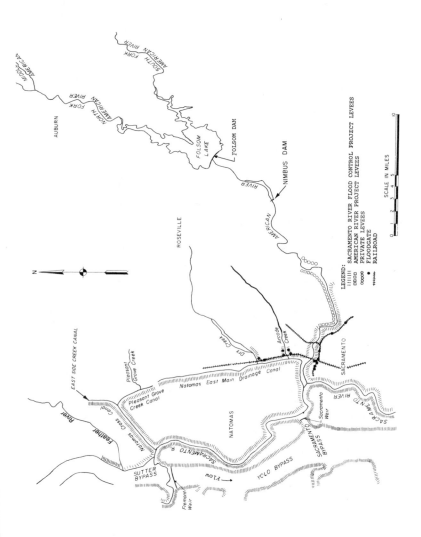

FIGURE 1.2 Existing flood control features of the American River Watershed. SOURCE: Sacramento District, USACE, 1991.

swampland reclamation districts upon petition of one or more property owners. For almost 50 years, flood control was in the hands of local landowners (Kelley, 1989)[2]:

> The Green Act . . . completely atomized flood control planning and construction down to the individual reclamation district. The Jeffersonian passion for localism, and for putting people on their own, had been entirely satisfied. The result was that for most of the next half-century, the Sacramento Valley would be scissored into a crazy-quilt of small reclamation districts whose levees followed property lines, not the Valley's natural drainage pattern. Flood control anarchy, and therefore massive flood control failure, would be the result.

In the absence of cooperative approaches to respond to the common flood hazard, each property owner, drainage district, and municipality historically sought to protect itself with levees to deflect floodwaters on its neighbors. In the 1870s, this precipitated a "levee building spiral" in which "each project responded to each other's threat by building further upstream and thus outflanking the other side . . ." (Kelley, 1989).

There was no statutory or judicial remedy for affected parties to prevent this from happening. Water law, such as it was, regarded rivers as a "common enemy" to be resisted by each property owner and town as best they could, regardless of consequences to each other. In fact, the Green Act authorized unilateral efforts by property owners to protect themselves. In 1876 a private landowner, Levi Moulton, sued another owner, William Parks, to prevent him from rebuilding and enlarging a dam/levee that threatened to raise and retain floodwaters on Moulton's land (*Moulton v. Parks*, 64 Cal. 166, 30 p. 613 (1883))[3]. Although the structure had been erected under authority of the Green Act, the local court granted a permanent injunction against the rebuilding of Parks's structure. This was upheld by the California Supreme Court in 1883 and set a precedent for judicial scrutiny of the reasonableness of piecemeal flood control measures in California.

Hydraulic mining at the headwaters of Sacramento River tributaries, which had begun in the 1850s, also contributed to the confusion. Miners washed away overburden to reach gold-bearing gravel, thereby clogging stream channels with debris, endangering navigation, and aggravating flooding. Mining interests ex-

---

[2]The definitive history of flood control in the Sacramento River basin is Robert Kelley's, *Battling the Inland Sea: American Political Culture, Public Policy, and the Sacramento Valley 1850—1986.* University of California Press, Berkeley, 1989. The committee is indebted to Martin Reuss, USACE senior historian, for his presentation and paper, "History of Flood Control in the Sacramento Valley" in which he summarized this complex history, drawing on Kelley's seminal study.

[3]The California Supreme Court actually decided for the plaintiff on the narrow ground that the Sutter County Board of Supervisors had no power to approve impoundment of floodwaters in another county, where the plaintiff's land was located.

erted such political power within the state that few limits were imposed by statute or court decision until a catastrophic flood struck in 1881. The flood prompted a series of lawsuits by property owners, perhaps encouraged by the *Moulton* case, against the mining companies. In 1884 the Federal Appeals Court for the 9th Circuit concluded that hydraulic mining was doing widespread damage and was a destructive public and private nuisance that must be halted (*Woodruff v. North Bloomfield Gravel Mining Co.*, 18F.753 (9th Circuit, 1884)).

Throughout the history of the Sacramento-American river flood control saga, the issue recurs as to what should be the geographical basis for action. The Green Act encouraged flood protection based on property boundaries, not hydrologic units. Most nineteenth-century levees were constructed by municipalities, landowners, or districts composed of groups of landowners. Seldom was cooperation achieved among private owners or districts sharing a watershed or facing each other across a common stream. Nor was flood control planning integrated with other functions of water resources management until the 1930s.

Gradually, as individual and collective landowners' flood control projects failed to stem the tide of flood damage and instead often shifted damage to other properties, more centralized institutions for flood management emerged. Two examples were the California Debris Commission created by Congress in 1893 to regulate hydraulic mining and the Sacramento Drainage District established by the state in 1905. Later in the twentieth century, authority was further concentrated under the federal Central Valley Project and the State Water Project. But local interests, such as those now represented by the Sacramento Area Flood Control Agency (SAFCA), continue to play a prominent role in advocating flood protection for particular communities and river reaches. With the advent of more centralized decisionmaking at the state and federal levels, flood control planning has increasingly been based on hydrologic rather than political boundaries. Implementation of plans, however, still depends to a considerable extent on local political support and local funds.

Competing Strategies of Flood Response

Throughout the history of Sacramento-American river settlement, competing engineering strategies for controlling floods have been advocated. The fundamental debate during the last three decades of the nineteenth century was between "levees only" and a combination of levees, bypass channels, and overflow basins. The former position was derived from the influential 1861 USACE report by Humphreys and Abbot that advocated "levees only" for ensuring navigation and flood control on the Mississippi River. Local sentiment in the Sacramento Valley, based on bitter experience, favored bypasses in addition to levees. The 1894 plan for Sacramento, called the Manson and Grunsky plan, developed under the California Commission of Public Works, advocated bypass channels and the widening of the Sacramento River. But the Dabney Commission in 1904, headed by a USACE officer, embraced the "levees only" doctrine.

A disastrous 1907 flood with a peak discharge of 600,000 cfs far surpassed prior estimates and discredited the "levees only" doctrine. The state in 1911 adopted a new plan by Thomas Jackson that incorporated levees plus bypasses and channel widening. Congress provided 50 percent federal funding to implement it in the 1917 Flood Control Act. Except for the lower Mississippi River basin, this was the first federal financial participation in flood control project construction prior to the 1936 Flood Control Act. With the addition of upstream storage after 1936 at Shasta, Folsom, Oroville, and other dams, paid for almost entirely with federal funds, the Jackson plan has been the blueprint for flood control in the Sacramento Valley.

Early approaches to flood control in the Sacramento/American River basin and elsewhere were entirely structural in nature. In the 1930s the National Resources Planning Board began to explore nonstructural alternatives to flood control, for example, conserving natural wetlands, land use planning (floodplain zoning), warning and evacuation systems, and financial mechanisms to offset the costs of flood losses. These types of approaches were strongly advocated in the 1966 report of the Task Force on Federal Flood Control Policy (U.S. Congress, 1966). The concepts of nonstructural floodplain management and flood insurance were adopted by Congress in the National Flood Insurance Act of 1968. But the debate over competing strategies continues, as evidenced by attitudes toward proposed new development in the Natomas Basin. While structural measures are unquestionably necessary in already developed areas, some argue that new development should be located and designed to avoid harm from floods without placing total reliance on structural protection measures.

Intergovernmental Cooperation and Cost Sharing

Recurring throughout the history of flood control in California, and throughout the United States, is the question of which level(s) of government should take initiative and bear the costs of achieving protection. Initially, in the absence of state or federal interest, costs were assumed by local communities, groups of landowners acting through a drainage district, or individuals. With the adoption of the Jackson plan in 1910, both the state of California and, in 1917, the federal government agreed to share the costs equally of building new levees, weirs, channels, and other facilities.

In 1935, Congress authorized USACE to build the Central Valley Project (see Box 1.1). This task was reassigned by Congress in 1937 to the Bureau of Reclamation. Thereafter in the Sacramento Valley, and across the United States, the federal government assumed the major share of the costs of building storage dams such as Shasta and Folsom. Nonfederal interests were required only to provide land, flood easements, and maintenance. The pendulum of cost bearing

BOX 1.1
THE CALIFORNIA CONTEXT

Flood control in California should not be treated in isolation—it should be seen as part of a complex system of water control and use that has evolved over a long period of time under the auspices of many government agencies. The needs of a growing population, and of agriculture in the fertile Central Valley, along with the lack of summer precipitation, result in tremendous water demands during the growing season, just when supplies are most scarce. At the same time, large storms and intense rainfall in the winter result in the need for flood protection, and the proximity to major population centers has spawned a growing demand for recreational water uses.

Two major systems—the Central Valley Project (CVP) operated by the Bureau of Reclamation, in conjunction with USACE, and the State Water Project (SWP) operated by the state—have evolved to control Sierra runoff. They support multiple goals and commitments, including water storage and transport for supply, flood control, recreation, power, navigation, and water quality purposes. Both the CVP and the SWP systems include major storage facilities in northern California and extrabasin transfers to southern California.

The CVP began as an emergency relief effort in 1935 and became one of the biggest projects in Bureau of Reclamation history. The linchpin of the project is Shasta Dam, completed in 1944, with a total capacity of 4.5 million acre feet. Other elements of the CVP include San Luis, Whiskeytown, Trinity, Folsom, and Friant dams, plus several major canals. This system relies on water stored in reservoirs such as Shasta, Trinity, and Folsom to replace water in the delta that is lifted at the Tracy pumping plant into the Delta-Mendota Canal. Because of its close proximity to the delta, low flow releases from Folsom are especially important for maintaining water quality in the delta.

The SWP includes several dams along the Feather River, with Oroville Reservoir as the primary facility. Oroville's primary purposes are water supply and flood control, although it also serves as a source of power for project operations and as a recreational resource. Water from winter rains and spring snowmelt is stored in Oroville and released during the summer and fall for irrigation and municipal uses. Water from Oroville flows down the Feather and into Sacramento River channels before entering the vast Sacramento—San Joaquin Delta. Huge pumps in the southern area of the delta pull water into the California Aqueduct, which distributes the water to contractors in the southern Central Valley and southern California or to pumps that lift the water for storage into San Luis Reservoir, which is jointly owned by the SWP and the CVP. But as demands on the system have increased, problems have arisen. For instance, because the delta is connected to San Francisco Bay and the Pacific Ocean, both CVP and SWP pumps can reduce fresh water inflow, allowing intrusion of saline water into the delta, and over time the water quality and dependent fish and wildlife within the delta have deteriorated.

The elaborate water supply system in California evolved in response to a variety of needs. As populations and demands on the resources have increased, a fundamental conflict has arisen in how to operate the reservoir system to obtain a balance between protecting the environment or mitigating environmental damage, ensuring adequate flood storage during the flood season, and providing adequate water and power supplies to meet the projects' contractual commitments.

thus swung almost entirely in the federal direction. Congress subsequently pared back the federal role. The Water Resources Development Act of 1986 expanded the nonfederal cost share for certain projects. The present situation on the lower American River is complex, with local interests that are acting through the Sacramento Area Flood Control Agency (SAFCA) taking primary responsibility for levee improvements, but thereby gaining credit toward the nonfederal share of possible construction of a new upstream storage project, which would remain predominantly a federal project.

Scientific Uncertainty

Two interrelated issues have plagued flood control efforts for the Sacramento-American river system. One is the question of how much protection should be provided to occupants and investments in floodplains. The other is how reliably we can estimate the level of protection afforded by an existing or proposed flood control project.

In the past, it was difficult to determine a sense of what would be an acceptable level of protection, since it was impossible even to estimate the risk of future extreme events. Empirical experience—the "flood of record"—provided the only guidance to levee builders. As each generation of levees was overwhelmed, the response was to build them higher, to stand up to a flood of the magnitude just experienced. But this approach failed to recognize the effects of human activities such as hydraulic mining on channel capacity. Floods in 1881 and 1907 far surpassed prior expectations in part because channels were clogged with debris. The "flood of record" approach also cannot accommodate the outlier natural event that exceeds recorded experience, especially in a region of short historical record such as the Sacramento Valley.

Since the development of modern statistical models for estimating peak discharges of extreme hydrologic events and hydraulic models for calculating the corresponding water levels, flood planners now can estimate the peak discharge, stage, and approximate geographic expanse of large floods that may not have occurred within the period of historical record. The Flood Insurance Rate Maps prepared by the National Flood Insurance Program are based on these techniques to determine areas subject to an annual chance of flooding of 1 percent ("100-year") and 0.2 percent ("500-year"). Floodplain management and mandatory purchase are required within the 1 percent flood zones. Yet, despite the appearance of precision, such estimates are still far from exact. The law is tolerant of scientific uncertainty and generally allows government the benefit of the doubt regarding floodplain management judgments (see Dingman and Platt, 1977; Platt, 1994).[4]

---

[4] In 1994, the U.S. Supreme Court in *Dolan v. City of Tigard* (No. 93-518, 62 *U.S. Law Week* 4576) held invalid a local requirement that a property owner dedicate a portion of her property that lay

---

BOX 1.2
WHAT DOES "100-YEAR FLOOD" MEAN?

The American people often hear references to a "100-year flood" but the meaning of the phrase is often unclear. As typically used, "100-year flood" means a flood that has a 1 percent chance of being equaled or exceeded in any given year. It has a 26 percent chance of occurring over the life of a 30-year mortgage. The terminology used to describe the 100-year flood can be confusing. The terms 100-year flood, 100-year recurrence interval flood, 100-year frequency flood, 1 percent flood, 1 percent annual chance flood, and base flood, which all refer to the same event, are often used interchangeably. Confusion can result because the 100-year flood is usually the only type people hear about, even though larger and smaller floods are likely to occur.

As commonly applied, the concepts of a 100-year flood and 100-year floodplain can be misleading. Technically, only the outer edge of a 100-year floodplain has a risk of 1 percent. The risk rises for sites closer to a river, ocean, or other water feature, and also at lower elevations, yet most people think of the entire area between the water body and the outer edge of the 100-year floodplain as subject to the same risk. Variation of risk is not usually shown on floodplain maps. There are areas within the mapped 100-year floodplain that may flood more frequently and to greater depths than others.

Uncertainties surround 100-year discharges and elevations, and mapping 100-year floodplain boundaries is at best an imperfect science. Estimates of the 100-year flood discharge (or flow rate) can be based on a range of techniques, and current techniques provide estimates that could be off as much as 5 to 45 percent (Burkham, 1978). Factors such as the size of the watershed, the availability and length of stream-gaging records, and the level of detail of mapping for use in determining model parameters contribute to the uncertainty in a 100-year flood discharge estimate. Flood discharges associated with infrequent events, such as the 500-year flood discharge, are more difficult to predict and have more uncertainty associated with them. Even if a fairly accurate 100-year discharge is determined, it may subsequently change owing to land use changes in the watershed and natural and human changes to the channel and floodplain.

SOURCE: Interagency Floodplain Management Task Force, 1992b.

Such calculations of course must be revised in light of actual experience. Thus the estimated level of protection provided by Folsom Dam and downstream levees on the American River was revised downward from the 100-year event to about a 70-year event after the 1986 flood. (See Box 1.2 for an explanation of the term "100-year flood.") Estimates of future rare events also may be affected by uncertainty resulting from climate change and land use change in the watershed.

---

within a floodplain plus an additional strip for a bikeway. The Court, however, did not question the method of determining the extent of the floodplain nor the need to limit development in such areas. The issue narrowly related to the requirement that the owner dedicate such areas to public use and access without compensation.

Folsom and Auburn Dams

Another element of the historical context that plays a part in understanding the current debate over flood control in the American River basin is the role of dams in the system. The flow of the American River upstream of Sacramento is regulated by Folsom Dam, a 340-foot-high, concrete-earthfill multipurpose structure completed by USACE in 1956 and operated today by the Bureau of Reclamation as part of the Central Valley Project. Folsom regulates runoff from about 1,860 square miles, receiving drainage from all three forks of the American River. Its maximum storage capacity is about one million acre-feet, of which 400,000 acre-feet is allocated to flood storage during the fall and winter months (Figure 1.3). Beyond the portion reserved for flood storage, the reservoir pool is allocated to power, irrigation, water supply, recreation, and releases to maintain minimum flows in the lower American River. Lower American River flows are also regulated by Nimbus Dam, a small regulating structure just downstream of Folsom Dam.

Together with an auxiliary dam and eight dikes, Folsom Dam impounds a reservoir with a shoreline of about 75 miles and a maximum surface area of some 12,000 acres. The nearby Sacramento Metropolitan Area, with a 1990 population of 1.48 million (up from 848,000 in 1970), makes heavy use of Folsom Lake as a recreational resource. The 18,000-acre Folsom Lake State Recreation Area is the most heavily used year-round facility in the state park system, with average annual user-days exceeding 3.4 million (Water Education Foundation, 1988; USACE, Sacramento District, 1991).

When Folsom Dam was planned in 1949, it was designed to protect against a flood characterized by a peak inflow rate of 340,000 cfs (680,000 acre-feet per day) and a 6-day inflow volume of 978,000 acre-feet, which at the time was thought to be a 500-year storm. The 6-day inflow (978,000 acre-feet) was about 2.4 times the size of the flood pool (400,000 acre-feet). Under these conditions, maximum releases would be 115,000 cfs (230,000 acre-feet per day), the standard to which the downstream levees were designed (U.S. Bureau of Reclamation, 1986). A series of floods in 1955, 1963, 1964, and 1986 radically changed the understanding of Folsom's estimated level of protection. As discussed in Chapter 2, the flood protection estimated to be provided by Folsom as currently designed and operated was subsequently downgraded to about a 70-year flood, a flood with a 1.4 percent chance of occurrence in any year (SAFCA, 1993).

In 1965 a second, larger dam was authorized by P.L. 89-161 to be constructed about 12 miles upstream from Folsom Dam near the town of Auburn. The proposed Auburn dam would have impounded runoff from the North and Middle Forks, controlling 973 square miles of the American River watershed and creating a two-pronged lake about 25 miles long. The originally proposed Auburn dam would have been another multipurpose structure, a concrete arch dam twice the height of Folsom (653 feet from base to crest) with a potential storage

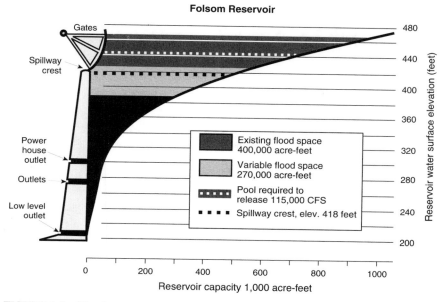

FIGURE 1.3   Flood storage space at Folsom Reservoir. SOURCE: Murray, Burns and Kienlen, 1993.

capacity of 2.3 million acre-feet, more than double that of Folsom. The full pool would have occupied 10,000 acres, and a total of 42,000 acres of land were scheduled to be acquired for the project.

Construction of the originally proposed Auburn dam by the Bureau of Reclamation began in 1967, despite strong opposition. A diversion tunnel and cofferdam to carry the American River past the construction site were completed in 1972. Work on the dam stopped in 1975, however, when an earthquake registering 5.7 on the Richter scale occurred near Oroville, about 45 miles north of Auburn. Subsequent study revealed a fault near the Auburn site. Some evidence suggested that the newly completed Oroville Dam may have triggered the earthquake, and the Auburn dam was put on hold indefinitely by the Bureau of Reclamation. About one-third of a billion dollars was invested at the Auburn dam site and average maintenance costs for the site amount to $1.5 million annually (USACE, Sacramento District, 1991).

Although the planned Auburn dam was redesigned to reduce seismic risk, the project as originally conceived lost support. According to USACE (USACE, Sacramento District, 1991), this was the result of two factors: (1) a 1986 change in federal policy concerning cost sharing of water development projects that would have raised the nonfederal share of the costs substantially and (2) more aggressive and effective opposition by environmental interests. The scenic and recreational values of the North and Middle Forks have indeed attracted wide-spread opposition to a permanent impoundment at Auburn. Ironically, the acquisition of over 30,000 acres for the Auburn dam impoundment actually helped to consolidate opposition to its completion. This area is now operated as the Auburn State Recreation Area and is heavily used for white-water rafting, camping, fishing, and hiking.

Reevaluation of the American River flood risk following the 1986 flood inevitably reopened the question of whether an Auburn dam should be built. SAFCA and other flood protection advocates offered a dry dam as a compromise alternative to the full-pool, multipurpose dam. As proposed, this dry dam would be used for flood storage only when needed; "frequency of impoundment" would depend on its design. No water would be permanently impounded, and the recreational use of upstream canyons would be largely unaffected except for impacts to valley walls and vegetation caused by occasional inundation. While considerably smaller than the originally proposed multipurpose Auburn dam, it would be the largest dry dam in the United States, and it has added an additional layer of controversy to this already complex decisionmaking process. The issues to be resolved include not only whether the dam is necessary and cost-effective, but whether the dam should have gates to control flow or remain ungated to discourage its conversion to multipurpose use. The committee shares complete consensus, however, that a dry dam of the size proposed for the Auburn site requires the safety margin and flexibility afforded by operational gates.

THE USACE PROJECT PLANNING AND DECISIONMAKING PROCESS

To have a full understanding of the American River flood control planning process, some familiarity with the USACE planning process in general is helpful. USACE studies for individual project planning move through a highly structured process that begins with a congressional study authorization, requires congressional and presidential approval, and ends (if successful) with project implementation (see Box 1.3). USACE planning is expected to provide technical analysis of the merits of different alternatives and the recommended plan to support informed decisionmaking at the local level (where the project will be implemented), in the executive branch, and in Congress.

The USACE district office has the primary responsibility for all aspects of project planning. After receiving congressional authorization to conduct a study, a district office is provided with a budget and assigned responsibility for recommending a plan for implementation, or recommending that no action be taken. In executing these responsibilities, the district office follows detailed planning procedures mandated by USACE Washington, D.C., headquarters. In addition, the district is expected to subject its planning to the myriad requirements of federal and state laws, such as the National Environmental Policy Act of 1969 (NEPA), the Clean Water Act, and the Endangered Species Act. Compliance with these various acts is reported in the study and, if appropriate, in an environmental impact statement (EIS) filed under NEPA.

In response to NEPA and similar legislation, by the mid-1970s USACE had introduced expanded public participation efforts in planning and made efforts to recognize the concerns of a broader array of interests. The district now is expected to solicit advice, and perhaps request particular technical studies, from other federal and state agencies. Extensive public participation is expected, often through formal public hearings at certain steps in the planning process. All of this external advice is expected not only to meet a legal requirement for consultation under different laws, but also to direct the study process and the resulting recommended plan of action.

Indeed, there were many procedural and substantive planning requirements in the various laws passed during the 1970s to provide a foundation for legal and political challenges to USACE planning and recommended plans. Over time USACE critics focused on environmental concerns have succeeded in slowing and then reversing the growth of the federal water development program. By the late 1970s the program had come to a near halt—no new construction projects being authorized—largely because of a congressional impasse over cost-sharing issues and other differences between the administrative and legislative branches over water planning. The program was restarted only after passage of the Water Resources Development Act of 1986 (WRDA, 1986).

WRDA 1986 is best recognized for dramatically increasing the required payment for the costs of USACE projects by nonfederal interests who benefit from the projects. For example, prior to 1986 the beneficiaries of a local flood control project would be expected to provide only the lands, easements, and rights-of-way necessary for the proposed project to be implemented. A major flood control reservoir required no local contribution. After 1986, cash payments were required in addition to the lands, easements, and rights-of-way requirement. Nonfederal costs could rise quite high, so high in fact that the law capped the nonfederal contribution at 50 percent of total costs, a substantial increase over the pre-1986 situation.

Another significant change was the requirement that the costs of feasibility studies be shared as well. Prior to 1986, study costs were a full federal responsibility. With WRDA 1986 the initial study is paid at full federal cost, but the costs of feasibility studies must be shared. For example, a nonfederal sponsor paid 50 percent of the costs of the 1991 American River Watershed Investigation.

These cost sharing requirements have put pressure on USACE to open its planning and decisionmaking to even more scrutiny than in the past. Those who pay for a study demand a greater say in all phases of the study process, and, as project implementation costs rise, the demands for influence on the recommended plan also increase. As a consequence of the recent challenges to USACE projects and of WRDA 1986, the USACE planning process not only is increasingly open

---

BOX 1.3
THE SIX STEPS OF THE USACE PLANNING PROCESS

USACE has been involved in development and management of the nation's water resources since 1824. The agency has planned and built projects to improve river navigation, reduce flood damage, and control beach erosion; it also has projects that generate hydropower, provide water supplies to cities and industries, regulate development in navigable waters, and provide recreational opportunities. In all, USACE manages nearly 1,500 water projects.

The planning process currently used to evaluate potential new projects was set out in the Water Resources Development Act of 1986 (P.L. 99-662), which establishes a framework for a cost-sharing partnership between the federal government and nonfederal interests and gives nonfederal participants a key role in project planning. According to USACE, there are six essential steps in the planning, design, and implementation of civil works projects:

1. Problem perception. The local community and/or a local government perceive a problem such as flooding or shore erosion that is beyond the local community's capabilities to alleviate.
2. Request for federal action. Local officials ask about USACE programs that might help; some small projects and technical assistance can be accomplished without congressional authorization.
3. Study problem and report preparation. The relevant district office is assigned to conduct a reconnaissance study, funded by the federal government. If a full feasibility study is warranted, the local sponsor must agree to share costs. This phase includes public involvement, including review of the draft feasibility report and draft environmental impact statement. The study follows the guidelines set out under the U.S. Water Resources Council's (1983) Economic and Environmental Principles and Guidelines for Water and Related Land Resources Implementation Studies. The study results are submitted to the USACE division office.

---

to environmental and other interests, but it also is a joint product of USACE and a local sponsor (such as SAFCA).

It was the degree of openness (or perceived lack of openness) of this planning process for the American River that provided the opportunity for critics to challenge the analysis of the Sacramento District and the plan preferred by the local sponsor. The fact that these challenges were made suggests that, although the process was open to inspection and comment after it was completed, it did not provide opportunity for significant, early input or fully incorporate the concerns of the interests who challenged the study. Of course, opposition may materialize no matter how open the planning process may be, but early identification of disagreements typically increases the opportunities for resolution.

4. Report review and approval. The division office reviews activities during the planning phase and provides technical review of the final feasibility report and EIS. This report is submitted to the Washington Level Review Center, which issues a public notice inviting comments and conducts a Washington level review. The final EIS is then filed with EPA and made available to the public, while the proposed final report is sent to heads of federal agencies and governors of affected states for comment. After considering comments, the Board of Engineers for Rivers and Harbors then submits recommendations to the Chief of Engineers. The Chief of Engineers considers all comments and prepares a final report for the Secretary of the Army; this report is reviewed by the Assistant Secretary of the Army (Civil Works) and the Office of Management and Budget, and then transmitted to Congress.

5. Congressional authorization. The Chief of Engineers reports are referred to the Committee on Public Works and Transportation in the House and the Committee on Environment and Public Works in the Senate; civil works projects are normally authorized by the Water Resources Development Act (Omnibus Bill) following hearings; occasionally, a USACE proposal is authorized by separate legislation or as part of another bill.

6. Project implementation. New projects are included in the President's budget based on national priorities and other factors; budget recommendations are based on the willingness of nonfederal sponsors to provide their share of the project cost. Funds for new starts are typically provided in the annual Energy and Water Development Appropriations Act. Project construction is managed by USACE, but done by private contractors. Most projects are operated and maintained by nonfederal sponsors, but where there is a need for continuing federal financing of project operation and maintenance, congressional appropriations are required.

2

Identification and Evaluation of Alternatives

In the 1991 American River Watershed Investigation (ARWI), the Sacramento District presented various alternative plans to provide flood control to Sacramento, including supporting analysis (USACE, Sacramento District, 1991). For each alternative plan, the 1991 ARWI provided estimates of the cost, expected benefits, and net benefits; the level of protection; and the environmental impacts and proposed environmental mitigation. Formal decisionmaking on the alternative plans was then based on these estimates.

In the USACE's planning process, the benefit-cost ratio is calculated to screen out inefficient alternative plans, as plans with negative net benefits are not eligible for federal funding. The alternative plan with the highest expected net benefits, consistent with applicable environmental laws and regulations, is designated the National Economic Development plan (NED) and is generally the plan recommended by the federal government. In the American River case, the NED plan included construction of a dam and 894,000-acre-foot reservoir at a site near Auburn. However local interests, as represented by the Sacramento Area Flood Control Agency (SAFCA), preferred a plan featuring a smaller dam and after consultation a plan including a smaller structure, offering, a 200-year rather than 400-year level of protection, became the selected plan.

During review of the 1991 ARWI by federal and state agencies and by public interest groups, concern about a number of technical issues emerged. These issues played some role in the rejection of the selected plan by Congress in 1992 and ultimately led to the creation of this committee. In a more recent document, the 1994 Alternatives Report (USACE, Sacramento District, 1994a) the Sacramento District presented a revised set of alternative plans, including estimates of costs and benefits. Unfortunately, the analysis supporting those new estimates is not scheduled for release until July 1995. In preparing the 1994 Alternatives Report, the Sacramento District had the opportunity to benefit from the technical debate that was generated by the 1991 ARWI and from interactions with this committee and many other parties. In addition, the 1994 Alternatives Report previewed the first application of USACE's new approach to evaluating flood control projects, an approach based on risk and uncertainty analysis.

This chapter discusses the development of alternative plans and the technical analysis used to estimate costs, benefits, and levels of protection. Subsequent chapters consider the analysis of environmental impacts and the new USACE approach to risk and uncertainty analysis. The committee's consideration of these issues was based largely on written and oral information provided by USACE, SAFCA and its consultants, and various critics of the 1991 ARWI. The committee was able to make firm recommendations on a number of technical issues, but many issues remain unresolved owing to lack of data and to the fact that the supporting technical analysis is not yet available. This latter fact has proven particularly problematic. Information related to that future document, received informally during briefings, indicates that the analysis supporting the 1994 Alternative Report is significantly different in many crucial respects from that which supported the 1991 ARWI. But the committee did not have formal written documentation of the analysis, and in most cases was uncomfortable about commenting on oral presentations and the few supporting documents that were available.

SELECTION OF PROJECT ALTERNATIVES

Perhaps the most critical step in the development of a flood control project is the selection of alternatives that will receive detailed analysis. Regardless of the potential effectiveness of a particular alternative, if it is not identified, it will not be selected. Furthermore, if popular alternatives are not selected for detailed analysis, it may be difficult to win support for the selected alternative, regardless of the potential effectiveness of the popular choices. Thus, this section looks specifically at the selection of alternatives in the American River planning process. (Additional discussion of the selection of alternatives and project planning in general is found in Chapter 6.)

Flood Control Measures

In developing project alternatives, USACE begins by identifying flood control measures that can be used alone or in combination. In the 1991 ARWI, the Sacramento District identified 23 flood hazard reduction measures, 13 pertaining to the main stem of the American River and 10 pertaining to Natomas. Of the 13 main stem measures, 4 were retained for further consideration and incorporated into flood protection alternative plans: (1) structural modifications to Folsom Dam to increase outlet efficiency; (2) increased downstream channel capacity to allow greater flood releases (so-called "objective releases") from Folsom Reservoir; (3) increased allocation of storage space in Folsom Reservoir to flood control; and (4) construction of a dam upstream of Folsom Reservoir (at Auburn). In the 1994 Alternatives Report, which excluded consideration of the Natomas Basin, the Sacramento District presented 17 measures, 8 of which were retained for further consideration. The latter included 4 measures for increasing the outlet efficiency of Folsom Dam, in addition to measures for increasing downstream channel capacity, increased flood control storage space in Folsom Reservoir, construction of a dam at Auburn, and raising of Folsom Dam and its spillway. The 1991 and 1994 flood control measures are summarized in Table 2.1.

Flood Control Alternative Plans

In the 1991 ARWI, the 4 surviving flood control measures were bundled into 6 alternative plans. Two alternatives were based on construction of a flood control dam at Auburn. Two other alternatives combined increasing flood control storage and outlet efficiency at Folsom with increasing downstream flow capacity. The fifth alternative was based solely on increasing the downstream channel capacity. The final alternative was based solely on increasing the proportion of flood control storage in Folsom Reservoir.

Seven alternative plans were presented in the 1994 Alternatives Report. Three of these were based on construction of a flood control dam at Auburn. Three other alternatives combined increasing flood control storage and outlet efficiency at Folsom with increasing downstream flow capacity. The final alternative combined increasing flood control storage and outlet efficiency at Folsom, without increasing the downstream flow capacity.

The alternative plans presented in the 1991 ARWI and 1994 Alternatives Report are summarized in Table 2.2, along with the estimated levels of protection and ratios of the net benefits to the net benefits of the NED plan. Note that the methods that the Sacramento District used to estimate the levels of protection in 1991 differed from those used in 1994; hence the estimates are not strictly comparable.

Criticisms of the 1991 Measures and Alternatives

The measures and alternatives presented in the 1991 ARWI were criticized on a number of grounds. Many of these criticisms focused on the evaluations of the alternatives; these are addressed in subsequent sections. However, some of the criticisms had to do with the perceived failure of the Sacramento District to consider and evaluate potentially effective alternatives. The most serious criticisms focused on Folsom Reservoir. In particular, critics argued that the district

TABLE 2.1 American River Flood Control Measures (Excluding Natomas)

Measure	1991 Report[a] Listed/Retained	1994 Preproject[b] Condition	1994 Report[c] Listed/Retained
Increased Outlet Efficiency of Folsom Dam and Reservoir			
Normalized use of auxiliary spillway	No	No	Yes/No
Structural modifications			
Lower main spillway	Yes/Yes	No	Yes/Yes
Enlarged river outlets	No	No	Yes/Yes
New river outlets	No	No	Yes/Yes
New tunnel outlets	No	No	Yes/No[d]
Conjunctive use of river outlets and main spillway (without modifying outlets)	No	No	Yes/No
Use of existing diversion tunnel	No	No	Yes/No
Improved flood forecasting and reservoir operation	Yes/No	No	Yes/No
Increased Flood Releases from Folsom Reservoir			
Levee/channel modifications	Yes/Yes	Yes	Yes/Yes
Setback levees	Yes/No	No	Yes/No
Flood control bypass south of Sacramento (Deer Creek)	Yes/No	No	Yes/No
Increased Flood Storage in the American River Basin			
Flood detention at Auburn	Yes/Yes	No	Yes/Yes
Existing upstream reservoirs	Yes/No	No	Yes/No
Multiple small-detention reservoirs	Yes/No	No	Yes/No
Offstream storage near Folsom	Yes/No	No	No
Out-of-basin storage on Deer Creek	Yes/No	No	Yes/No
Increased flood space in Folsom	Yes/Yes	Yes	Yes/Yes
Raised Folsom Dam and spillway	Yes/No	No	Yes/No[d]
Other Measures			
Divert flood flows into Sacramento River deep water ship channel	Yes/No	No	No
Miscellaneous nonstructural	Yes/No	No	No

[a]Measures listed for consideration in the 1991 American River Watershed Investigation, Sacramento District, U.S. Army Corps of Engineers.

[b]Measures from the 1991 ARWI that were treated as part of the pre-project condition (i.e., measures already or planned to be implemented) in the 1994 Alternatives Report, Sacramento District, U.S. Army Corps of Engineers.

[c]Measures listed for consideration in the 1994 Alternatives Report, Sacramento District, U.S. Army Corps of Engineers.

[d]Measures that may be reconsidered before final recommendations are made.

TABLE 2.2 American River Flood Control Alternative Plans

Alternative	Level of Protection[a] (years)	Net Benefits/ NED Net Benefits[b]
1991 ARWI		
Auburn Dam—894,000 acre-feet	400	1.0
Auburn Dam—545,000 acre-feet	200	0.80
Folsom Modification and Reoperation (1)	150	0.56
Increase maximum Folsom flood		
control storage to 650,000 acre-feet		
Lower Folsom spillway		
Increase objective release to 130,000 cfs		
Folsom Modification and Reoperation (2)	100	0.30
Increase maximum Folsom flood		
control storage to 470,000 acre-feet		
Lower Folsom spillway		
Increase objective release to 130,000 cfs		
Levee Modification		
Increase objective release to 145,000 cfs	100	0.30
Increased Folsom Flood Storage		
Maximum flood control		
storage—590,000 acre-feet	100	0.34
1994 Alternatives Report		
Auburn Dam—894,000 acre-feet	455	1.0
Auburn Dam—545,000 acre-feet	270	0.70
Auburn Dam—380,000 acre-feet	200	0.30
Folsom Modification and Reoperation (3)	244	0.24
Modify Folsom outlet works		
Increase objective release to 180,000 cfs		
Folsom Modification and Reoperation (4)	217	0.21
Variable Folsom flood control storage		
450/670,000 acre-feet		
Modify Folsom outlet works		
Increase objective release to 145,000 cfs		
Folsom Modification and Reoperation (5)	185	0.19
Variable Folsom flood control storage		
475/670,000 acre-feet		
Modify Folsom outlet works		
Increase objective release to 130,000 cfs		
Folsom Modification and Reoperation (6)	152	0.32
Variable Folsom flood control storage		
495/670,000 acre-feet		
Modify Folsom outlet works		
Maintain objective release at 115,000 cfs		

[a]Level of protection was computed differently in 1991 and 1994.

[b]For the 1991 ARWI, the divisor is the net expected benefit for the 1991 NED plan; for the 1994 Alternatives Report, the divisor is the net expected benefit for the 1994 NED plan.

failed to adequately consider modification of the operation of Folsom Dam, which, coupled with improvements in the dam's outlet capacity, might significantly increase the effectiveness of the existing storage. Some of these criticisms were addressed in the 1994 Alternatives Report. Most notable is a reoperation plan for Folsom Reservoir that will increase the winter flood control space based on the availability of storage space in the three largest reservoirs in the upper American River basin. This plan is expected to be implemented independently of the ongoing planning process and hence is considered an existing condition in the 1994 Alternatives Report.

Issues of Importance in the 1991 and 1994 Alternative Plans

In considering the alternative flood control plans in both the 1991 and 1994 reports, the committee elected to focus on four specific elements: use of Folsom Reservoir, the question of gates in the Auburn Dam alternatives, the Deer Creek alternative, and nonstructural measures.

Folsom Reservoir

As noted above, the 1991 ARWI was criticized for failing to give sufficient consideration to ways to maximize the flood mitigation potential of Folsom Reservoir, including the use of flood forecasts. How valid is that criticism? Before addressing this question, consider how the operation of Folsom Reservoir determines its effectiveness at reducing flood risk in Sacramento.

Folsom Reservoir provides the primary means of reducing flood flow in the lower American River. The flood reduction potential of the reservoir depends on the amount of water that can be stored as compared to the difference between the amount that enters the reservoir during major flood events and the amount that can be safely released. At full pool, Folsom Reservoir has a storage capacity of about one million acre-feet. But Folsom is a multipurpose reservoir; in addition to flood control, its purposes are water supply, hydropower, and recreation. Unfortunately, there are conflicts among these objectives. If the reservoir were to be operated for an assured water supply alone, the optimal strategy would be to keep the reservoir as full as possible. If the reservoir were to be operated for flood control alone, the optimal strategy would be to keep the reservoir as empty as possible. Clearly, the reservoir cannot be operated to maximize both of these objectives simultaneously.

One solution to this dilemma is to allocate storage amounts separately to flood control and water supply. Nominally, the top 400,000 acre-feet of storage space in Folsom Reservoir is allocated to flood control; the remainder is allocated for water supply. This allocation is not rigid, however, owing to the timing of flood events in the watershed. Potentially damaging floods occur only during the winter storm season, which lasts from the beginning of November through the end of March. Hence the full flood storage pool need be available only during this period. The manner in which the flood storage space is managed is specified by a flood control diagram that was originally formulated in 1956 and modified in 1977 and 1987. Under the 1987 diagram (USACE, 1987), the flood control storage space must be increased from zero on October 1 to a maximum of 400,000 acre-feet on November 17, at which level it must be maintained until February 8. Between February 8 and May 31 the flood control space is to be varied according to the accumulated seasonal precipitation, which is closely related to the depth of snowpack in the upper American River watershed. This currently used approach to managing the flood control space in Folsom Reservoir could be modified to improve flood control effectiveness (as is being considered with the Folsom reoperation, discussed below). Such improvements may or may not come at the expense of water supply or other water resources purposes (see Chapter 6 for additional discussion).

The seasonal allocation of flood storage determines the amount of storage available for flood control prior to a flood. The effectiveness of the available storage depends on how it is used during a flood event. Obviously, it is desirable to release water as rapidly as possible without causing downstream damage during a flood, since that frees up storage space in the reservoir. But there are constraints on how rapidly water can and should be released. First, there are physical limitations on the maximum discharge rate from the reservoir. Folsom Reservoir is severely limited in this regard. For example, the primary flood-release structures, the five main spillway bays, cannot discharge water at the objective release rate of 115,000 cfs until the flood control storage has been filled to about half of total capacity. (The objective release rate is the design discharge capacity of the channel and levee system downstream of the reservoir; sustained flows in excess of this rate could cause levee failure.) Second, there are administrative and legal limitations on releases. The 1987 *Water Control Manual for Folsom Reservoir* (USACE, 1987) provides that as an operating guide, "releases from Folsom Dam shall not be increased more than 15,000 cfs or decreased more than 10,000 cfs during any 2 hour period. . ." This limit on the rate of increase of discharge rates (the so-called "ramping rate") is intended to minimize bank sloughing and caving downstream and to allow time to prevent downstream loss of life and damage to property. The 1987 Water Control Manual also limits the maximum controlled release to 115,000 cfs, up until the time at which the storage level of the reservoir reaches full pool. At full pool the release policy is governed by an emergency spillway release diagram that is designed to protect the reservoir from failure due to overtopping. There is one additional constraint that is applied to the operation of the reservoir during floods: while inflows are rising, the controlled discharge from the reservoir cannot exceed the inflow rate. This requirement ensures that in no flood event will the peak discharge below the reservoir exceed the peak discharge into the reservoir. Note that this is a de facto policy that is not explicitly specified in the 1987 Water Control Manual.

All of the above constraints on the operation of Folsom Reservoir can be modified to some extent. Changing the physical constraints, of course, requires structural modifications to the reservoir and levees. The remaining constraints are administrative and legal and could be changed by appropriate agreements.

As noted above, the 1991 ARWI considered a number of measures for improving the flood control effectiveness of Folsom Reservoir, including lowering the main spillway, using flood forecasting to draw down Folsom Reservoir in advance of a potentially severe storm, increasing the objective release, increasing the allocated flood space in Folsom, use of storage in upstream reservoirs, and raising Folsom Dam. Of these, the use of flood forecasting, use of storage in upstream reservoirs, and raising Folsom Dam were not incorporated into any of the proposed alternatives. In the 1994 Alternatives Report, the original 1991 measures were reconsidered, although increasing the Folsom flood space in accordance with the amount of water stored in upstream reservoirs (Folsom reoperation) was considered to be a without- project condition. New measures in 1994 included construction of new outlet works, as well as altered use of the existing outlet works. As in 1991, measures involving flood forecasting and the raising of Folsom Dam were not incorporated into alternatives, although apparently the latter measure is still being considered.

It is clear that the Sacramento District considered a number of strategies for increasing the flood control effectiveness of Folsom Reservoir. The most notable of these is the Folsom reoperation, which is considered a without-project condition in the 1994 Alternatives report. Also relevant is the decision by the Sacramento District to reject use of flood forecasts, as well as some other approaches to Folsom operation.

Folsom Reoperation

One measure considered in the 1991 ARWI was increasing the Folsom flood control storage allocation to 650,000 acre-feet. This measure was included with lowering the Folsom spillway and increasing the objective releases in an alternative that provided an estimated 150-year level of protection. The lost water supply resulting from the increased flood control allocation was computed to cost about $10 million per year, or about 20 percent of the total annual cost of the alternative. Subsequently it was realized that if the Folsom pool were lowered in accordance with the water stored in the largest upstream reservoirs, the expansion of the flood pool would not necessarily represent a loss to water supply. On the basis of this realization, several potential operating rules were considered; of these, the so-called "670 plan" became a without-project condition in the 1994 Alternatives Report. Under this plan, the flood control space in Folsom Reservoir would vary between 400,000 and 670,000 acre-feet, based on the day of the year and the reservoir storage space available in the French Meadows, Hell Hole, and Union Valley reservoirs. Between December 1 and March 1, the Folsom

TABLE 2.3 Estimated Volume of Water That Must Be
Stored in Order to Control the Flood of the Given
Recurrence Interval to the Given Objective Release

| Recurrence Interval (years) | Required Volume (1,000 acre-feet) | |
	Objective Release of 115,000 cfs	Objective Release of 180,000 cfs
100	498	232
200	770	452
400	1,115	748

NOTE: The volume estimates are based on the USACE flood quantile estimates for the 3- and 5-day floods, without the expected probability correction, and on the design hydrograph used in the 1991 ARWI, without any adjustments for upstream storage.

flood control space would be maintained at 400,000 acre-feet if the empty space in the three upstream reservoirs totaled at least 200,000 acre-feet. Any incremental reduction in the upstream space would require a corresponding incremental increase in Folsom's flood space. When all of the empty space in the upstream reservoirs was filled, the flood-storage space at Folsom would be maintained at 670,000 acre-feet (SAFCA, 1994a). Although Folsom reoperation was considered a without-project condition in the 1994 Alternatives Report, it still must be approved prior to its adoption.

This proposed modification of the operation of Folsom Reservoir represents a significant increase in the flood control effectiveness of the reservoir. An idea of the relative magnitude of this increase can be obtained from Table 2.3, which gives for different levels of protection the volume of water that must be controlled if the corresponding flood peak is to be kept from exceeding an objective release of either 115,000 or 180,000 cfs. The table was developed by computing the area enclosed above the objective release and below the design hydrograph for the given recurrence interval. It is based on the design hydrographs used in the 1991 ARWI, without the expected probability correction. From Table 2.3 it can be seen that the maximum additional storage of 270,000 acre-feet provided by the proposed modification represents about 35 percent of the volume required to control the 200-year event to 115,000 cfs. For the 400-year events, the amount is 24 percent.

Flood Forecasting and Flood Control Effectiveness

In both the 1991 ARWI and the 1994 Alternatives Report, the Sacramento District considered and then rejected a measure involving the use of weather forecasts to draw down Folsom Reservoir in advance of a storm. This decision was based on the conclusion that weather forecasting was not sufficiently accurate. The committee also doubts the efficacy of early releases, given the current limitations of precipitation and runoff forecasting, physical and administrative limits on pre-flood-peak release rates from Folsom, and the fact that Folsom reoperation will enable use of about 70 percent of the available storage space in the reservoir. The committee thinks, however, that forecasting may be of value in devising strategies for regulating floods that exceed the Folsom flood pool capacity so as to minimize the amount by which the actual Folsom outflows exceed the objective release. In addition, dam operation decisions that clearly take available forecast information into account are more likely to be acceptable to both the dam operators and the public than decisions that do not make use of all available information. The committee recommends, therefore, that the Sacramento District, the Bureau of Reclamation, and the state of California keep abreast of developments in precipitation forecasting and develop the capability to exploit major improvements in forecasting accuracy.

Folsom Operation During Flood Events

As previously discussed, maximum flood-reduction effectiveness requires rapid discharge of water during a flood event. In this regard, Folsom Reservoir presents three issues: limitations in the outlet structures at Folsom; appropriateness of the rules governing the release of water from Folsom during floods; and actual operation of the reservoirs during past floods.

During a flood event, Folsom releases water over the main spillway, through river outlets in the spillway, and through the power penstocks. The main spillway has eight gated bays. Five of these bays discharge down the spillway into a stilling basin at the base of the dam; they constitute the main release mechanism. The river outlets were designed to operate concurrently with the five main spillway bays. The remaining three spillway bays, called the auxiliary spillway bays, discharge to a flip-bucket energy dissipator. These bays were designed to help pass water during extreme floods to protect the dam against overtopping.

Unfortunately, the existing outlet facilities are inadequate and limit the flood control effectiveness of Folsom Reservoir. When the pool is at the bottom of the current flood space (400,000 acre-feet of storage), the five main spillway bays can pass only 6,500 cfs. At a flood storage space of 500,000 acre-feet, the main bays cannot pass any water. The original operation of Folsom Reservoir depended on the concurrent use of the river outlets and the five main spillway gates. Shortly after the dam became operational, however, it was discovered that concurrent use caused cavitation damage to the spillway. Even with subsequent modifications to the river gates, concurrent operation of the river and spillway gates has been avoided.

These limitations on flow releases severely constrain the current operation of Folsom and would be especially constraining under the proposed reoperation.

For this reason, several of the proposed measures involve construction of new outlet structures. In addition, Countryman (1993) made a number of recommendations for improving the efficiency of Folsom Reservoir with the existing structures. These include concurrent operation of the river outlets and five main spillway gates and use of the three auxiliary spillway gates during normal flood operations. Countryman calculated that use of his "maximum outlet plan" would increase the releases during the FEMA 100-year flood by over 60,000 acre-feet. This represents about 8 percent of the volume required to control the 200-year flood to 115,000 cfs (Table 2.3). Although this is not a large percentage, given the low level of protection currently provided Sacramento, the recommendations of Countryman (1993) should be considered seriously. The committee was told that the main spillway gates and the river outlets are assumed to operate concurrently in the analysis supporting the 1994 Alternatives Report.

The committee did not attempt to evaluate in detail the appropriateness of the ramping rates or of the de facto requirement that outflows be less than inflows during the period of increasing inflow. The committee was told that in the analysis supporting the 1994 Alternatives Report the ramping rates were increased by 33 percent for flow up to 25,000 cfs and increased by 100 percent for flows above 25,000 cfs. Operating with these new rates would improve the flood-reduction effectiveness of the reservoir. The committee conducted its own analysis of the increases in water levels and velocities associated with the ramping rates. The results of this analysis show no reason why ramping rates must be held at 15,000 cfs per 2 hours. The committee recommends that the Bureau of Reclamation and the Sacramento District consider the impacts of operating Folsom with higher ramping rates.

The more critical issue is the way the reservoir is actually operated in practice. Up to the present, the operator has had to compute reservoir inflows on the basis of observed increases in water levels. This problem alone results in a 4-hour delay in releases. It is the committee's understanding that the flow measurement issue is being remedied by the installation of telemetering equipment at flow monitoring stations in the three main upstream tributaries. The committee strongly supports the development of real-time capacity for monitoring inflows to Folsom Reservoir and of a means for accurately gaging outflows from Folsom and Nimbus reservoirs.

Another important operational problem is the failure of operators to follow the rules. In its discussion of the 1986 operation of Folsom Dam, the Bureau of Reclamation stated that prescribed rule curve operation should be viewed as "hypothetical." The agency goes on to say (Bureau of Reclamation, 1986)

> operators are reluctant to rapidly increase the volume of outflow and consequently affect the floodplain unless such increases are clearly warranted. It is estimated that actual operating efficiencies, when compared to hypothetical operation, are about 80 percent.

If this statement is true, then either a change is needed in the constraints placed on operators with clearer specification and formulation of release rules, or else planning assumptions should be revised. This issue is explored further in Box 2.1. In addition, the operation of Folsom during the 1986 flood, discussed later in Box 2.2, illustrates these issues.

Special training and use of forecasts are two approaches that could be used to improve the performance of reservoir operators during flood events. Forecast-based rules, if well thought out and tested in advance, can to lead to better decisions than can be made on an ad-hoc basis under emergency conditions. Special training in use of the operating rules could make effective use of simulation exercises, in which operators develop experience in decisionmaking under both historical and hypothetical extreme events. Simulation exercises can prepare operators to take those actions early in a storm that are required to reserve flood storage to control very large events.

Recommendations on Folsom Operations

Folsom Reservoir is the critical component in the flood control system for Sacramento. Consequently, it is essential that it be operated as efficiently as possible during floods. Based on the 1986 flood experience, it is clear that there were problems with how Folsom was operated: the ramping rates were excessively conservative, needed gages were not installed, operators were not careful about retaining flood control storage, and the dam did not go on alert when the rest of the state did. (See Box 2.2 and Figure 2.1.) Since 1986, several changes have been made or proposed, including new operating rules in 1987 and a proposed reoperation plan. But, in spite of these changes, the committee was uncertain about the current and future operating efficiency of Folsom Reservoir. The reasons for this uncertainty include the following:

• The Folsom Flood Management Plan, referred to in the 1994 Alternatives report, was not completed in time for committee inspection. This plan is intended to "maximize the flood control capability within the existing 400,000 acre-foot flood control reservation of Folsom and improve the stream-gage network and flood-forecast system for the American River basin upstream from the reservoir" (USACE, Sacramento District, 1994a). The committee was not provided any details on the recommendations in this report and hence unable to evaluate their potential effectiveness.

• The proposed Folsom reoperation was not final at the time of this report, and represents a major change in reservoir operation. It is notable that the Folsom Flood Management Plan apparently does not consider Folsom reoperation.

• The current ramping rates may be unduly conservative, as recognized in the 1994 Alternatives Report.

BOX 2.1
PLANNING ASSUMPTIONS AND
OPERATIONAL EFFECTIVENESS

There is at times a "disconnect" between the operating assumptions made in planning and what actually happens during the operation of a flood control facility. Sometimes such differences are caused by how mechanical systems perform, and if deficiencies in system operation are identified, operating policies should be revised to reflect actual conditions or the deficiencies should be corrected. (The 1991 ARWI, for instance, accepted many operating constraints as givens, while the Sacramento District's upcoming Supplemental Information Report is said to be more aggressive in questioning these restraints and attempting to find ways to eliminate problems.) One example of an operating constraint is the reluctance of operators to allow free flow over the Folsom spillway because it is difficult to estimate the resultant release from lake levels; this technical problem could be solved by installing stream gages or other flow-measuring devices.

Sometimes such differences are caused by the nature of the assumptions. Conservative assumptions, indicating low expectations of operational efficiency, might under-represent the time it would take operators to respond to new information or neglect the availability of real-time forecasts that might allow operators to anticipate reservoir inflows. On the other hand, optimistic assumptions might fail to consider that in major floods, rain and streamflow gages can fail, communications lines can break, general confusion can impair the decisionmaking process and the communication of decisions to operating personnel, and gates or other structures can fail to operate as anticipated. For instance, there is a long history of problems with erosion of spillways affecting operations, including the 1983 flood at Glen Canyon Dam.

A key issue is whether operators follow the rules based on the planning assumptions. For instance, there is significant concern as to whether system operators will implement drastic release policies, which might cause damage to the floodway or even put lives at risk, early in a flood. As an example, serious problems were encountered in the operation of Painted Rock Dam on the Gila River in Arizona (Rezac, 1993). On January 20 of 1993, heavy rains in the region resulted in

---

• The potential addition of new outlet works will require fresh consideration of the operating policy at Folsom.

• The potential construction of a dam at Auburn will require a new operating policy for Folsom, both to maximize efficiency of the combined system and to minimize environmental impacts.

It is important to stress that while the committee was uncertain about the current and future operating efficiency of Folsom Reservoir, it did not believe that these uncertainties were sufficiently large to compromise the validity of the 1994 Alternatives Report. Hence the committee does not suggest that these uncertainties must be resolved before a flood control alternative is selected for Sacramento.

In any case, whatever the decision regarding flood management for Sacra-

major flood flows that filled the reservoir to 60 percent of capacity and the reservoir release schedule called for a release of 22,500 cfs. However, USACE dam operators received approval to hold releases to 12,500 cfs because higher releases would cause extensive downstream flooding and bridge closures. Subsequently, the reservoir filled and uncontrolled releases over the spillway reached 25,600 cfs. In the February storm of 1986 on the American River, dam operators delayed making large releases because of fears of downstream damage, and were reluctant to increase releases during night hours (Bureau of Reclamation, 1986).

Water control plans often adopt low expectations for system operations and set rules and regulations accordingly, thereby institutionalizing less that the most effective use of such facilities. On the other hand, optimistic assumptions are likely to result in underestimation of the actual flood risk in a basin, result in false assurances of safety, and divert attention away from the critical processes where improvements could reduce flood risk. USACE has observed (USACE, 1959),

> The temptation to infringe on flood control space is sometimes strong, because usually losses to other functions are obvious, and losses to flood control (although usually much greater) may not occur or indeed probably will not occur in any particular case. Consequently, a very rigid attitude against infringement on flood control space must be maintained at all times.

> This issue is of national importance. Flood control operating assumptions used in planning, and the corresponding water control plan and its associated rules and regulations, need to balance the level of operation that can be achieved with the level of operation that will actually be realized. In some cases effort needs to be directed at developing the institutional and legal framework needed to ensure that operators will actually follow the agreed upon flood control rules and regulations in the face of opposition by powerful special interest groups, frightened citizens and their political representatives, and their own reluctance to cause damage and possibly risk loss-of-life. Moreover, few operators are mentally prepared to spill large quantities of water because of a possible event that is beyond their experience, particularly after they have spent years managing and conserving water to enhance hydropower, water supply, and environmental needs.

mento, there needs to be serious evaluation of the operation of Folsom Reservoir. This evaluation should be a cooperative effort, involving the Bureau of Reclamation, USACE, the state of California, and the U.S. Weather Service and should include:

• consideration of technological capabilities in precipitation and runoff forecasting, remote sensing of rainfall, real-time monitoring of upstream reservoir storages, soil moisture, snowpack, and streamflows, and rainfall-runoff simulation;

• consideration of operating rules that exploit current technological capabilities, including rules governing reservoir operations when the flood capacity is exceeded;

• quantitative assessment of various operating rules;

BOX 2.2
OPERATION OF FOLSOM DAM IN THE 1986 FLOOD

The February 1986 flood in the American River basin was a timely warning. It renewed attention to the flood risk to people and property in the floodplain, while causing relatively little damage. The flood risk in Sacramento results from the extensive development within the floodplain and the hydrologic risk presented by large flows from the American River. Attempts have been made to reduce this risk by providing flood storage in Folsom Lake and an extensive levee network along the lower American River through Sacramento. Of great concern is the recognition after the 1986 flood that the estimated level of protection provided by the existing system was perhaps 60 to 70 years, rather than the previous estimate of 120 years.

This reappraisal gave the appearance to some that it was revision of the operation of Folsom that caused the change. Moreover, there were public pronouncements and discussion in the media indicating that the near-disaster was caused by the failure of the Bureau of Reclamation to operate the facility correctly (Harris, 1986; Williams, 1993). Lessons learned from the 1986 event can be constructive to ongoing efforts to plan future flood operations, including the following:

1. Folsom Lake began the flood event with 710,000 acre-feet of water in storage, which corresponds to 100,000 acre-feet of encroachment within the 400,000 acre-feet of storage nominally reserved for winter flood regulation (Bureau of Reclamation, 1986) (shown in Figure 2.1). On February 4, 1986, the Sacramento District warned the Bureau of Reclamation by letter that encroachment in Folsom violated flood control regulations. However, regulations allowed that encroachment at the time of the mid-February storm because of the dry conditions that had prevailed in the watershed near the end of the flood season. But drought does not preclude flooding; state-wide flooding also marked the end of the 1976 to 1978 drought in California.

Lesson: Even when conditions have been relatively dry, a major flood can occur.

2. On February 13 and 14 the California Department of Water Resources (CDWR) began preparations for a full flood fight, given computer projections of a extraordinary storm approaching the state from across the Pacific (CDWR, 1986). The American River flood flows began in earnest on February 15, with inflows rising to over 60,000 cfs early the next day, but Figure 2.1 shows that Folsom operators did not begin to evacuate the flood control storage volume, nor did releases from Folsom match the inflows to the lake. Operators expressed a major concern for the effect of large Folsom releases on recreational facilities in the lower American River floodway; releases were held to 20,000 cfs for 36 hours. This is inconsistent with the 1977 USACE flood control diagram in force at the time; the diagram states that when Folsom storage is in the flood control reservation the water "shall be released as rapidly as possible" subject to ramping limits. Even after increased releases from Folsom began on February 16, and before they reached the 115,000-cfs limit, Folsom releases continued to lag behind inflows into Folsom Lake by 30,000 cfs or more. USACE-prescribed ramping limits of "15,000 cfs during any 2-hour period" do not appear to have limited the rate of increase of Folsom releases during the 1986 flood, nor were physical release rate limits at Folsom Dam a constraint given the initial elevation of the reservoir.

Lesson: Procedures need to be adopted to ensure that flood releases are made as required by operating regulations if intended flood risk reduction is to be achieved.

3. Folsom operations were primarily based on the actual inflow to Folsom Reservoir calculated from lake level changes (Figure 2.1). This calculation ignored the accumulation of water in the cofferdam near the Auburn dam site above Folsom. Written operating procedures do not mention this accumulation of water. Because this cofferdam was designed to breach with the 30-year flood flow, its accumulation distorted the effective inflow to the Folsom-cofferdam system and the accumulated storage in the two reservoirs, which ended up in Folsom Reservoir when the cofferdam finally breached.

Lesson: Plans need to be updated to reflect changes in facilities in basins and "temporary" structures.

4. The Bureau of Reclamation lacked a forecasting system that effectively routed flows from rainfall through the upstream hydropower reservoirs, and the cofferdam near Auburn and into Folsom Lake. Thus, the operators did not make the timely releases that were warranted given that: the storage level in Folsom Lake was above the flood control reservation diagram; the flood storage reservation for Folsom had begun to increase after February 14 to the original 400,00 acre-feet; the "storm of the season" had been forecasted; there was additional risk due to the accumulation of water in the Auburn coffer dam, which was designed to fail in a 30-year flood.

Lesson: Operating procedures should reflect storage levels and the general regional risk of flooding, not a single forecast, even if large flood flows have not yet occurred.

If the Bureau of Reclamation had been able to more closely match outflow to inflows while inflows were less than 115,000 cfs, then releases into the American River would not have exceeded 115,000 cfs during the 1986 flood using the nominal storage capacity of the reservoir, even without anticipation of the Auburn cofferdam failure. Fortunately, disaster was averted by the use of extra surcharge storage in Folsom and by the ability of the downstream channel and levee system to handle releases of 130,000 cfs. Lessons drawn from the 1986 experience should not be forgotten.

- justification of constraints on release rates; and
- operator training and other means of improving operator performance, including use of continuous interactive simulation of storm events.

The evaluation of reservoir operating rules should be an ongoing process, so as to reflect changes in technology and in the physical system. Furthermore, there is a national need for assessment and monitoring of the effectiveness of operating rules at all major reservoirs with flood control obligations. This need

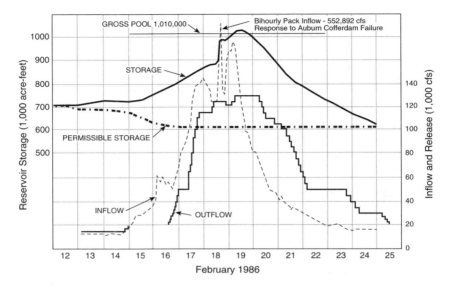

FIGURE 2.1 Flood operation of Folsom Lake during the February 1986 flood. Note that at the start of the flood, Folsom Lake was encroached within the 400,000 acre-feet nominally reserved for winter flood regulation. SOURCE: Bureau of Reclamation, 1986.

arises from a number of factors, including potential changes in flood regime due to changes in climate and watershed conditions and to changes in political and economic demands on reservoir storage space.

Gated Auburn Dam

In the 1991 ARWI, the Sacramento District was criticized because it included gated outlet structures in its preliminary designs of the proposed flood control dams at Auburn. The argument against gates was that a gated flood control reservoir could be converted in the future into a multipurpose reservoir. In the 1994 Alternatives Report, the District held firm on its inclusion of gates, and even increased the number. The committee fully concurs with this decision on the grounds that gates are essential to safety and flood control efficiency and, as discussed later, because gates allow operational flexibility that can be used to minimize environmental impacts.

The proposed 425-foot dam at Auburn (the 1991 "recommended" alternative) would be well over twice the height of any ungated dams constructed by USACE. As with any dam of this height, there would be uncertainties regarding potential cavitation damage to the main spillways and scour at the downstream toe. If such damage occurred during a major flood, gates would make it possible to reduce flows to less damaging levels, greatly reducing the possibility of dam failure. For this reason alone, it would be extremely unwise to construct a 425-foot-high dam without gates. The ability to control flows during a flood event would also facilitate emergency actions in the American River downstream of the dam and at Folsom Reservoir. Finally, gated outlet works make it possible to operate the dam conjunctively with Folsom Dam, potentially providing both improved flood control efficiency and reduced environmental damage in the canyon.

Offstream Flood Control Storage on Deer Creek

In the 1991 ARWI, the Sacramento District considered a measure involving diversion of American River flood flows to a detention basin in the Deer Creek watershed. On the basis of preliminary calculations, this measure was determined to be very costly compared with other measures and hence was dropped from further consideration. Subsequently, the Sacramento District developed a conceptual design for a Deer Creek project that alone would be able to provide Sacramento with a 200-year level of protection (USACE, Sacramento District, 1994b). The project provides for diversion of American River flood flows from Folsom Reservoir via a connecting channel to a detention basin in the Deer Creek watershed about 10 miles south of Folsom Reservoir. Design of the Deer Creek project assumes that the seasonal flood control storage in Folsom Reservoir will remain at 400,000 acre-feet. Releases would be made from Folsom Reservoir to the Deer Creek detention basin only after it had been determined that the American River had achieved the objective release of 115,000 cfs from Folsom Reservoir. During nonflood periods, no water would be stored in the Deer Creek detention basin. (Apparently, no water supply objective for the Deer Creek reservoir was proposed or included in the Sacramento District's investigation.) Estimated capital costs for the project range from approximately $1.2 billion to $1.8 billion depending on project design. This represents approximately $2,500 per acre foot of storage. This can be compared to the cost of $67 per acre foot of storage for the 1991 NED Auburn Dam alternative.

The August 1994 draft analysis of the proposed project (USACE, Sacramento District, 1994b) concluded that the project is technically feasible, but expressed serious doubt about the social feasibility. The 1994 Alternatives Report repeated similar conclusions. The report indicated that there are major concerns about potential environmental impacts, specifically with respect to rare and endangered species, that would require expensive coordination and consultation with the U.S. Fish and Wildlife Service. In addition, the report indicated that there would be significant land use conflicts and high land costs, which would range from $50,000 to $200,000 per acre in some areas. The report also noted that construction of a 300-foot-wide channel connecting Folsom Reservoir to the proposed detention site on Deer Creek would pass through the middle of a number of developments and would have a severe effect on the overall development plan for the area. The report concluded that because of the lack of "social feasibility" of the Deer Creek project further study of offstream flood control storage on Deer Creek should be discontinued.

The Sacramento District's analysis apparently only investigated the potential for a 600,000-acre-foot storage reservoir that would not receive any water from Folsom Reservoir until the objective release of 115,000 cfs from Folsom Reservoir had been reached. The study did not report on the possibility of including a smaller Deer Creek reservoir together with a combination of other measures in order to produce an overall package of flood control measures to provide 200-year protection to Sacramento. It may very well be that such a package would not be competitive with the alternatives that were retained in the 1994 Alternatives Report, and that USACE analysts were able to reach that conclusion on the basis of their analysis of the full Deer Creek project. If so, discussion of this conclusion in the 1994 Alternatives Report would have forestalled potential criticism.

Nonstructural Measures

The flood protection alternatives considered in the 1991 ARWI and the 1994 Alternatives Report consist largely of structural measures (e.g., reservoir storage, levee improvement, increased channel conveyance). Nonstructural measures, including floodplain zoning, relocation, flood warning, floodproofing, minimum elevation building design, mandatory insurance, and evacuation capabilities received little consideration. This omission of nonstructural measures is discussed in more detail in Chapter 5. The committee believes that nonstructural measures can make a significant contribution to flood damage reduction, especially to flood damage reduction in currently undeveloped areas such as Natomas. Therefore, the committee recommends that nonstructural flood damage reduction measures be evaluated together with the structural measures for implementation in the American River watershed.

FLOOD RISK REDUCTION FROM ALTERNATIVE PLANS

Once alternative plans have been developed, they must be evaluated carefully. The most critical performance criterion of a given plan is its expected net benefits, the difference between the expected benefits of the plan and its costs. For a flood control project, the expected benefits consist mainly of the difference between the expected value of flood damages with and without the project. At any location, the expected value of flood damages is the integral over all possible flood stages of the product of the flood damage that would occur at a given stage and the probability of that stage. To evaluate a given project, it is necessary to estimate and total the expected value of flood damages at all locations subject to flood damage, with and without the project. This requires a complicated set of interrelated calculations, based on statistical, hydrologic, hydraulic, geotechnical, and economic models. The analysis framework that the Sacramento District used for these calculations is the "design event" method.

A design event is a hypothetical flood that produces a unique set of flows and stages, and hence damages, throughout the project area. The design floods are characterized by a single measure of event magnitude (such as peak flow or maximum 3-day volume) and a corresponding exceedance probability. It is assumed that at all locations in the project area subject to flood damage, the exceedance probability of the flow, stage, and corresponding damage is equal to that of the design event magnitude. Hence in the case of a 100-year design flood, the damage produced at all locations in the project area is assumed to have an annual exceedance probability equal to 0.01.

The application of the design event method to the American River is some-what complicated. The starting point in the analysis is Folsom Reservoir. On the basis of a long-term streamflow record from the gaging station just downstream of the reservoir and the record of storage changes in Folsom Reservoir, the Sacramento District estimated the inflows to Folsom Reservoir and the probabil-ity distribution of rain-flood inflow volumes for various durations. For each exceedance probability considered, the District then constructed a "balanced" design hydrograph based on the corresponding flood inflow volumes. For an alternative without upstream storage, each inflow design hydrograph was routed through Folsom Reservoir. For an alternative with an upstream reservoir, the Folsom inflow design hydrograph was separated into two components reflecting the inflows to the upstream site and the inflows from the remainder of the Folsom drainage basin. The upstream inflows were routed through the upstream reser-voir and the resultant upstream outflow hydrograph was recombined with the inflows from the remainder of the Folsom drainage basin. The recombined inflow hydrograph was then routed through Folsom Reservoir.

The discharge from Folsom Reservoir was then augmented to account for the additional drainage area between Folsom Dam and downstream locations. The additional discharge was determined from a rainfall runoff model of the contrib-uting drainage areas, based on a design storm with the corresponding exceedance probability. Hydraulic analysis was then used to determine stage hydrographs for the design event. For the lowest part of the river, the hydraulic analysis was particularly complex because of the complicating effects of water levels in the Sacramento River and in the bypass system. Next, the Sacramento District esti-mated the damage associated with each design event. This was based largely on the stage and flow hydrographs at critical locations where it was expected that levee failure would first occur. Finally, the Sacramento District estimated the expected value of flood damages for each alternative, including a without-project alternative, by integrating the product of damages and the corresponding exceedance probabilities.

The design event concept has been used for well over 50 years to design engineering works involving peak flows. The classical application of this concept is the design of storm drainage systems in urban areas. Typically, the area of interest is small, the hydrology can be characterized by a single parameter such as maximum rainfall depth accumulated over the time of concentration over the watershed, and there is a one-to-one correspondence (such as the rational formula) between hydrologic inputs and outputs. In more general applications of the design event concept to a particular flow system, it is assumed that the maximum flows at all points of interest in the system can be expressed adequately in terms of one-to-one correspondences between the flow and a single numerical parameter called the design event magnitude. Under this assumption, the peak flows at all points of interest have the same exceedance probability as the corresponding design event magnitude.

Over the years, however, design event methods have been applied to increasingly more complicated design problems, in which many factors affect the flows within the system. For example, in reservoir storage systems, the design event magnitude may be characterized by the maximum 3-day inflow volume, but other factors relating to the time distribution and shape of the inflow hydrograph may have important effects on the reservoir outflows and on flows at critical points downstream. Under these conditions, the downstream flows are not determined solely by the design event magnitude, and the assumption that downstream flow exceedance probabilities are equal is no longer valid.

Nonetheless, to avoid complexity in the risk and expected damage computations, designers have tended to retain the use of the event-magnitude exceedance probability and to adopt conservative fixed values of the secondary factors. For example, the so-called "balanced" design hydrograph used in the 1991 ARWI studies is synthesized by assuming that the maximum 1-, 3- and 4-day volumes under the hydrograph all have the same specified exceedance probability as determined from the flood volume frequency curves; other hydrograph shapes and combinations of hydrograph volumes and probabilities are ignored for simplicity. The so-called "operational contingencies" used by USACE are other examples of such assumptions. These assumptions are made to protect the public at risk by providing some additional margin of safety. But simplifying assumptions, if overly conservative, can lead to upwardly biased flood risk estimates, and in turn to inefficient projects. In the case of contentious projects, such as flood control for Sacramento, the conservative assumptions also can be lightning rods for criticism.

Consider the design situation on the American River. The design events are a set of "balanced" inflow hydrographs to Folsom Reservoir, each with an assumed exceedance probability. But the probability distribution of peak flood discharges at downstream locations depends on a number of factors, including the time distribution and shape of the inflow hydrograph (in addition to its magnitude), initial encroachment of Folsom Reservoir, actual reservoir operating decisions, contributions of downstream tributaries, concurrent flows and levels in the Sacramento River and bypass system, and the flows and water levels at which levees fail. In applying a design event method to the American River, the Sacramento District made a number of assumptions about these factors based on the engineering judgment of its analysts.

In order to evaluate the appropriateness of any of these assumptions, it is necessary to determine whether other reasonable assumptions would have resulted in significantly different answers. In the 1991 ARWI, the results of sensitivity analyses were presented for several of the operational contingency assumptions. The new methods that USACE is developing to evaluate risk and uncertainty may reduce the need to make conservative assumptions about some of the critical factors, such as the flows and levels at which levees fail. Apparently some of the "contingency assumptions" made in the 1991 ARWI were handled through uncertainty analysis in the 1994 Alternatives Report. Unfortunately, the committee was not able to review the details of this analysis. Instead Chapter 4 gives a general evaluation of the new methods.

In considering the methods and assumptions used by the Sacramento District to estimate flood damages, the committee attempted to evaluate the significance of the assumptions, as well as comment on their reasonableness. Correctness, per se, was rarely the issue. The committee was not able, however, to do formal sensitivity analysis and hence was not always able to reach firm conclusions. In such cases the committee merely indicated that the particular assumptions warranted further investigation. The committee's evaluation focused on the most critical components of the flood damage estimation: (1) development of design hydrographs at Folsom for unregulated conditions (estimation of probabilities); (2) development of design hydrographs below Folsom (accounting for the effects of Folsom and Auburn reservoirs); (3) computation of stage hydrographs at critical damage locations; and (4) estimation of damages at critical damage locations (determining the probability at which levees fail). Generally speaking, these components correspond to hydrologic, hydraulic, and geotechnical modeling. Each of these modeling components is discussed below, with an emphasis on the methods and assumptions that have generated the most controversy and that the committee judged to be most critical. Most of the discussion focuses on the 1991 ARWI, since the 1994 Alternatives Report does not provide supporting technical documentation. Several critical issues that emerged during the committee's review of the USACE analysis are also discussed.

Development of Inflow Design Hydrographs for Unregulated Conditions

Development of the unregulated design hydrographs for use as inflow hydrographs for Folsom Reservoir is a key component of the design process, because it is here that probabilities are introduced into the process. There are two basic steps: estimation of the probability distribution of unregulated flood volumes and construction of unregulated hydrographs.

Estimation of the Probabilities of Unregulated Flood Volumes

Estimation of the probability distribution of unregulated flood volumes was based on analysis of stream gage data collected at Fair Oaks since 1905. These data were adjusted to remove the effects of storage in Folsom Reservoir and in five upstream reservoirs. Series of adjusted annual maximum flows, representing unregulated flows to Folsom Reservoir, were developed for durations of 1, 3, 5, 7, 10, 15, and 30 days, for both rainfall and spring snowmelt floods. Probability distributions for these series were estimated using the program REGFQ: Regional Frequency Computation (USACE, 1982), which was written at the Hydrologic Engineering Center based on a method described by Beard (1962). Only the estimated distribution for rainfall floods was used in subsequent analyses. (In cases where flooding is due to distinct climatic mechanisms, such as rainfall and snowmelt, it is prudent to separately analyze annual flood series from each mechanism. Because large floods on the American River never result from spring snowmelt events, the estimated probability distributions of the spring snowmelt volumes are not needed in the subsequent analysis.)

The program REGFQ estimates the parameters of the Pearson Type III distribution for the logarithms of flow using the method of moments. On the basis of examination of the REGFQ user's manual (USACE, 1982), it appears that the program is consistent with guidelines in Bulletin 15, published in 1967 by the Water Resources Council (WRC, 1967) to provide federal agencies with a uniform technique for estimating flood flow probabilities. REGFQ does not incorporate subsequent modifications to the recommended techniques, presented in Bulletin 17B (IACWD, 1982). The most significant of these modifications involve adjustment for historical floods, estimation of generalized skew, and testing and accounting for outliers. REGFQ also includes one feature that is not included in Bulletin 17B: adjustment of the log-space moments to ensure that for all probabilities of interest the corresponding d-day average flow is always a decreasing function of duration, d. This is done by developing smoothed relationships between log-standard deviation and log-mean and between log-skew and log-mean for the various durations considered.

Once probability distributions have been estimated for the various durations, REGFQ applies an expected probability adjustment to the estimated flow quantiles. This adjustment was developed by Beard (1960) to ensure that nationwide failure rate experience for statistically designed structures would be consistent with the failure probabilities adopted for the design. However, as discussed in Chapter 4, use of the expected probability adjustment does not yield unbiased estimates of the risk of flooding or expected damages.

To evaluate the significance of the various idiosyncrasies of the estimation procedure used by USACE, probability distributions of the Fair Oaks rain and flood data from 1907 through 1986 were estimated in a manner consistent with Bulletin 17B. In making the calculations, the lowest rain-flood data point was

TABLE 2.4 Comparison of Quantile Estimates for 1-Day and 3-Day Mean Flows of Annual Rain Floods

Recurrence Interval (years)	1-Day Quantile (1,000 cfs)			3-Day Quantile (1,000 cfs)		
	Bulletin 17B	USACE	USACE (without expected probability)	Bulletin 17B	USACE	USACE (without expected probability)
25	151	165	158	112	115	110
50	198	220	210	149	150	145
100	253	285	271	192	195	185
200	316	360	341	243	245	231
500	413	485	453	323	325	304

NOTE: The upper and lower 95-percent confidence limits for Bulletin 17B 100-year 1-day flows are 190,000 and 363,000 cfs, and for 100-year 3-day flows are 138,000 and 267,000 cfs.

found to be a low outlier; it was removed and the conditional probability adjustment was used. The resulting quantile estimates for the 1- and 3-day mean flow are shown in Table 2.4. For the 1-day flows, USACE estimates are higher than those based on Bulletin 17-B by 9 to 17 percent. For the 3-day flows, USACE estimates are virtually identical to those based on Bulletin 17B.

Also shown in Table 2.4 are USACE estimates without the expected probability correction. For the 1-day flows, about half of the difference between USACE's quantile estimates and those based on Bulletin 17B is due to the use of the expected probability correction. Most of the remaining difference is due to the application of the Bulletin 17B correction for low outliers. In the case of the 3-day flows, for which there were no outliers, USACE quantile estimates without the expected probability adjustment are 1 to 6 percent lower than those based on Bulletin 17B. It should be noted that the observed differences between the USACE and the Bulletin-17B estimated quantiles are much less than the uncertainties in the estimates.

Although the effect of the expected probability adjustment is small in relation to the uncertainty in the quantile estimates, the committee concluded that the Sacramento District should not have applied the expected probability adjustment. The purpose for which the adjustment was developed is not relevant in the Sacramento situation, and, as explained in Chapter 4, the adjustment yields biased estimates of level of protection and expected damages. It appears that the Sacramento District did not use the correction in its analysis supporting the 1994 Alternatives Report; however, as discussed in Chapter 4, the committee disagrees with the procedure the District did use in the 1994 Alternatives Report to estimate level of protection.

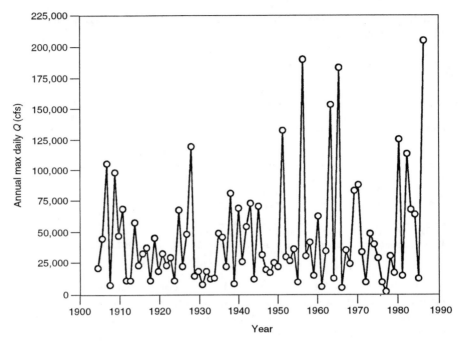

FIGURE 2.2 Annual maximum daily flow, American River at Fair Oaks, 1906-1986; adjusted for affects of regulation.

A critical issue in the estimation of probabilities of unregulated flood volumes is the apparent increase in the frequency of large floods in the Fair Oaks record since 1950 (Figure 2.2). The six largest annual maximum 1-day flood volumes in the Fair Oaks record (adjusted for Folsom effects) occur in or after 1950. This apparent increase in flood magnitudes has led to historically increasing estimates of the vulnerability of Sacramento to catastrophic flooding. It is clearly a very important issue, and yet the committee was unable to discover any scientific studies that explain the apparent increase in flood magnitudes. Later in the chapter the discussion returns to this issue.

Construction of Inflow Design Hydrographs—Unregulated Conditions

For each exceedance probability evaluated, an inflow design hydrograph was constructed that preserved the appropriate volumes. In the 1991 ARWI the duration of the design hydrograph is 4 days. Hence, for a given exceedance probability, the maximum 1- and 3-day volumes of the associated design hydrograph equal the 1- and 3-day volume quantiles with the same exceedance probability. Similarly, the total volume of the design hydrograph equals the 4-day volume, which was estimated by averaging the flow rate associated with the 3- and 5-day volumes. The general shape of the hydrograph is based on the probable maximum flood hydrograph developed in a 1980 study evaluating the adequacy of the Folsom spillway.

In the 1991 ARWI, the 100-year flood inflow hydrograph was reduced by 47,000 acre-feet to account for storage in upstream reservoirs. The reduction was made at a linear rate for the first day of the design hydrograph. No adjustment was made for floods greater than the 100-year flood, based on the assumption that no storage would be available for these extreme events.

The Sacramento District's correction for upstream storage has been criticized as unduly conservative (Williams and Galton, 1987; Swanson and Associates, 1992). On the basis of a compilation of data on the storage available in the largest upstream reservoirs 15 days prior to the annual maximum flood, Swanson and Associates (1992) estimated that the value of 47,000 acre-feet used by USACE has an exceedance probability of about 98 percent. (That is, in 98 percent of the years, the available storage exceeded 47,000 acre-feet.) The mean available storage was estimated to be 289,000 acre-feet, an amount equal to about 70 percent of the available flood storage at Folsom. Williams and Galton (1987) also criticized the fact that the Sacramento District adjusted only the first day of flow on the design hydrograph, arguing that because of the travel time involved the adjustment should have been made on the second, and most critical, day of the hydrograph. The Sacramento District's response to these criticisms is that the upstream reservoirs are not operated for flood control and cannot be counted on for storage in large floods. Furthermore, the reservoirs are located in the upper part of the watershed, capturing runoff from only 14 percent of the total watershed area.

The Sacramento District clearly made conservative assumptions in accounting for upstream storage. The most critical of these is that upstream storage would provide no benefits during events as rare as or rarer than the 1 percent event. What is the potential impact of this assumption, and was it too conservative? Consider again Table 2.3, which gives for various recurrence intervals the volume of storage required to control the design hydrograph to 115,000 or 180,000 cfs. The mean available upstream storage of 289,000 acre-feet is about 38 percent of the volume required to control the 200-year event, for an objective release of 115,000 cfs. Hence it is a significant amount of storage. However, the effective storage potential in these reservoirs is much less because of their upstream location. The reservoirs were designed to be drawn down to low levels in the fall and re-filled by spring snowmelt, not by winter rainfall floods. Although the available flow records are not easy to interpret, the reservoir storage records show only relatively minor storage increases, consistent with the 14-percent watershed area figure, for the major Folsom floods. In all cases the stored water was retained in the upstream reservoirs for at least several weeks after the flood. This is consistent with the use of the reservoirs for hydropower generation. It appears, therefore, that the upstream reservoirs have limited flood control value not because their available capacity is not effectively utilized but because they are located upstream from most of the significant rain-flood runoff-producing area. Furthermore, the operators of the upstream reservoirs have no responsibility for providing flood control for Sacramento. How can flood control credit be allocated to that storage when it is not managed for flood control? Fortunately, the Sacramento District has found an effective way to exploit the potential benefits of the upstream storage (USACE, 1993). This can be done by adjusting the Folsom flood control space in concert with the available storage in the upstream reservoirs, as specified in the Folsom reoperation plan discussed previously in this chapter.

Development of Design Hydrographs Below Folsom

Accounting for Effects of Folsom Dam

For events with an exceedance probability of greater than about 2 percent, the effects of Folsom storage were accounted for by directly using 32 years of flows measured at Fair Oaks. For rarer events, design hydrographs were routed through Folsom, based on several assumptions:

- initial flood control storage encroachment of 80,000 acre-feet;
- initial release of 20,000 cfs;
- outflow lags inflow by 4 hours;
- releases increased by a maximum of 7,500 cfs/hr and decreased by a maximum of 5,000 cfs/hr;
- for storage at or below main spillway, release based on full capacity of river outlets;
- for storage above main spillway, release based on main spillway, but not river outlets; and
- maximum surcharge storage of 50,000 acre-feet, as prescribed by the emergency spillway release diagram.

These assumptions were subject to considerable criticism. The assumed initial flood encroachment, which accounts for flood storage that is used by a lesser flood event preceding the design event, was criticized as being too large. The operating assumptions and the assumption of a maximum surcharge storage of 50,000 acre-feet were criticized as being too conservative.

The initial flood control storage encroachment accounts for the occurrence of a storm event in advance of the design event. It is another example of an uncertain factor about which ad hoc assumptions have to be made when using the design storm method. It should be noted from Table 2.3 that the assumed encroachment of 80,000 acre-feet is about 10 percent of the volume required to control the 200-year event to 115,000 cfs and about 7 percent of the volume required to control the 400-year event to the same flow. In the 1994 Alternatives Report, the Sacramento District accounted for the occurrence of an antecedent storm by using a two-wave design hydrograph. The committee found no evidence that would lead to the conclusion that either of these approaches was inappropriate.

In considering the assumptions in the 1991 ARWI having to do with routing of water through Folsom, the committee held to the belief that the assumptions used in the analysis of a reservoir should accurately reflect the operating rules of the reservoir. The actual operation of Folsom Reservoir during the 1986 flood event was not as efficient as the assumptions used in the 1991 ARWI. (For details, see Box 2.2 and Figure 2.1, earlier in this chapter.) After the 1986 event, modest improvements were made in the operation of Folsom Reservoir. More recently, the Bureau of Reclamation and USACE have been working on a management plan that would further improve the operational efficiencies of Folsom Reservoir. Given the state of flux in the operations of Folsom, the committee did not find the 1991 operational assumptions to be unreasonable.

Because the committee did not have documentation supporting the 1994 Alternatives Report, it was not able to examine in detail the methods used to evaluate the operational assumptions regarding Folsom Reservoir. It is the committee's understanding, however, that the assumptions do not reflect the improvements in operational efficiency that would be possible with telemetered discharge information and other operational changes, particularly with respect to the alternatives involving new outlet works. For example, the committee was told that the 1994 analysis assumed a 10-hour delay in initiating releases from Folsom prior to the second flood wave, if the flood reservation is evacuated after the first flood wave. A 10-hour delay seems to be too long, especially if telemetered flow data are available from upstream gages. Ten extra hours of flow at 115,000 cfs represents about 12 percent of the total volume required to control the 200-year flood to 115,000 cfs (Table 2.3). Hence this amount is large enough to warrant further consideration of the reasonableness of the assumption.

The committee was unable to evaluate the way in which USACE incorporated the Folsom reoperation plan in the analysis supporting the 1994 Alternatives Report. As previously discussed, this plan requires that the available flood storage in Folsom be adjusted in accordance with the status of storage in the French Meadows, Hell Hole, and Union Valley reservoirs. This dependence of Folsom operation on storage in upstream reservoirs is an example of the secondary factors that complicate the use of the design event concept. Apparently, USACE attempted to consider this particular factor in its new risk and uncertainty procedures, although the committee was not able to evaluate the way in which this was done.

Accounting for Effects of Auburn Alternatives

Design inflow hydrographs for Auburn alternatives were constructed by scaling the Folsom design inflow hydrographs by 67 percent, to account for the additional drainage area contributing to Folsom. The actual percentage used was based on drainage area ratios, analysis of normal annual precipitation, and historic flood flow data. The design hydrographs were routed through the Auburn alternatives; the resulting outflow hydrographs were then used as design inflow hydrographs to Folsom Reservoir. The committee was not aware of any criticisms of the District's approach to accounting for Auburn storage and did not investigate the issue in depth.

Accounting for Downstream Inflows

For each design event, discharge from Folsom Reservoir was augmented to account for the additional drainage area between Folsom Dam and downstream locations. The additional discharge was determined from a rainfall runoff model of the contributing drainage areas, based on a design storm with the corresponding exceedance probability. The committee was not aware of any criticisms of the Sacramento District's approach to this portion of the analysis and thus did not investigate the issue in depth.

Hydraulic Modeling

The next step in the analysis was to estimate stage hydrographs for critical locations on the lower American River. In the 1991 ARWI, this was done by using a one-dimensional gradually varied flow analysis. For the lowermost portion of the American River, which is affected by water levels in the Sacramento River, this analysis required assumptions of concurrent water surface elevations in that river. This section considers the appropriateness of those assumptions as well as the adequacy of the methods used in the hydraulic modeling.

Assumptions About Confluence of American and Sacramento Rivers

In modeling water levels in the portion of the American River affected by water levels in the Sacramento River, assumptions must be made about the magnitude of the flood hydrograph on the Sacramento River and its timing relative to that of the American River. With regard to timing, the Sacramento District used the same relative timing as occurred in the 1986 flood event. This meant that the peak of the American River hydrograph was assumed to occur about one-half day ahead of that of the Sacramento River. With respect to flood magnitude, the District assumed the occurrence of a 100-year event on the Sacramento River when modeling the 100-, 200-, and 400-year events on the American River.

The use of the 1986 flood event as a model for relative timing was criticized because the sluggish operation of Folsom Reservoir during that event delayed the arrival of the American River hydrograph (Williams and Galton, 1987). If Folsom had been operated so that outflows equaled inflows, the American River peak would have arrived at the junction about one and one-half days earlier. Swanson and Associates (1992) estimated that had this occurred the peak flow at the confluence would have been reduced by about 30,000 cfs (Swanson and Associates, 1992), a reduction of about 6 percent.

Assumptions about the level of the Sacramento River during a flood event on the American River are required by the event focus of the design flood method. The District clearly made conservative assumptions, but in the absence of a very elaborate analysis of the joint occurrences of floods on the two rivers (such as discussed by Dyhouse (1985) for case of the Missouri River), the assumptions are reasonable. As the Sacramento District demonstrated in the 1991 ARWI, flood peaks on the two rivers can occur within 1 or 2 days of each other. Furthermore, the hydrographs of both rivers are typically broad, so that a 1- or 2-day lag in peaks does not significantly affect the peak of the combined flows.

Hydraulic Models

In the analysis supporting the 1991 ARWI, the model HEC-2 was used to compute water levels at various locations on the lower American River. HEC-2 is a USACE model for computing water-surface profiles of one-dimensional, steady-state, gradually varied flows. In recognition of the unsteady nature of flood flows, USACE subsequently has developed a one-dimensional unsteady flow network model of the lower American River, based on the USACE model UNET. The committee supports this change in modeling approach. For levee overtopping or failure, the UNET model does not consider the momentum conservation between the river flow and the flow in the floodplain and thus the velocity and direction of flow in the floodplain are not properly calculated. In order to estimate the flood residual risk behind the levees, a two-dimensional unsteady flow model is needed to calculate the force and momentum of the flows to assess the possible damage, warning systems, floodproofing and evacuation. This two-dimensional model is also needed to calculate the flow behavior at the junction between two or more rivers such as the occurrence of flow separations downstream of the junction. These flow separations may cause the formation of eddies to cause excessive bank erosion. A one-dimensional model cannot predict flow separation.

During the 1986 flood, sediment accumulated upstream from the Fremont Weir up to about 1 foot above the weir crest, although some of this material probably was deposited over time in earlier years. This sediment deposit blocked the flow over the weir. Thus, this modeling effort should also consider the potential role of sediment accumulation at the various control structures, especially at the Fremont Weir.

Hydraulic modeling of the potential affects of a dam break due to earthquake should be investigated if construction of a new dam at the Auburn site is pursued because such a dam would be very close to an earthquake fault.

Geotechnical Analysis

Once flood peak flows and water levels were computed throughout the lower American River, the District estimated the damages that would occur for each of the design events considered and for all alternatives, including the without-project conditions. For alternatives involving increased objective releases, the process was more complicated. Before evaluating the damages associated with these alternatives, it was first necessary to design and cost measures for improving the channel and levee system to handle the increased objective releases. This is an extremely critical part of the analysis, since project costs are obviously an important factor in the decision process.

The objective releases investigated by the Sacramento District in the 1991 American River Watershed Investigation are 115,000, 130,000, 145,000, and 180,000 cfs. The feasibility of conveying these objective releases in the American River channel downstream from Folsom Dam is determined in part by the adequacy of the downstream levees and revetments to contain flows within the channel without failure.

Where the investigation showed that levees became unstable for flows above 130,000 cfs, the Sacramento District proposed to stabilize the levees with slurry cutoff walls. Flood releases over 165,000 cfs would overtop existing levees and require raising longer reaches of levee. With the higher objective releases, some areas upstream from the project levees would require new levees or floodwalls. Setback levees were also considered by the Sacramento District in its alternatives. These levees were considered for the lower American River in order to confine it to a narrow corridor.

A component of the 1991 ARWI and the 1994 Alternatives Report focused on determining what additional work would be necessary to the levees and channel revetments in order to allow conveyance of the objective releases with a reasonable degree of certainty. To answer this question, the Sacramento District investigated the possibility of levee failure by breaching, overtopping, seepage under the levee, and other causes.

The following sections contain an analysis of the methodologies and data used by the District in their levee investigations, together with some suggestions for improving the analysis. Further analysis of the geomorphology conditions in the American River basin that are relevant to the levee investigation are discussed in a later section.

The 1991 ARWI Analysis

As indicated in the 1991 ARWI, levees along the lower American River have been constructed and modified over many years. Near downtown Sacramento, the levees were originally designed to accommodate a peak flow of 180,000 cfs. Today, with the existence of Folsom Reservoir, flood flows can be attenuated for longer duration, but the levees cannot safely pass a sustained flow of 180,000 cfs. After the February 1986 flood, extensive geotechnical evaluation of the levees was conducted (USACE, Sacramento District, 1991, Appendix M). The ARWI concluded that there are reaches of levees that will exhibit structural deficiencies with sustained flows as low as 130,000 cfs. The ARWI concluded that levees along the lower river are believed to be able to safely accommodate a sustained flow of only 115,000 cfs.

The 1991 ARWI evaluation of levee failure on the lower American River concentrated on failure caused by encroachment on freeboard. The elevations at which the levees might fail were determined based on a projection of the impacts of various water levels on the physical system. Failure projections were based on varying degrees of encroachment, knowledge of levee conditions, exposure to high velocities or wave run-up and overtopping, and levee performance during the February 1986 flood. The analysis of levee failure was based on several factors, including:

• The assumption that the levee improvements described in the Sacramento River Urban Flood Control System Evaluation, Phase 1 (Sacramento Urban Area) would be complete (i.e., Sacramento area levees are stable up to their design flow).

• The observed condition of the levees in relation to geotechnical evaluation of the function of the system during the February 1986 high flows.

• Hydrologic observations and forecasts developed in the Hydrology Appendix to the 1991 ARWI.

On the basis of the these parameters and procedures, the 1991 ARWI developed an estimate of potential levee failures (Table 2.5) that specified the maximum flow or stage that could occur on a specific reach of levee before failure. This analysis was deterministic and was based primarily, if not entirely, on remaining freeboard. The 1991 ARWI also qualified this estimate of potential levee failure by indicating that these estimates are "for flood damage estimates only. Actual levee failures may occur at higher or lower flows and stages" (Table 2.5).

On the basis of this deterministic evaluation of levee reaches that would fail at varying flows and/or stages, the 1991 ARWI detailed a list of levee and channel modification projects (including necessary revetment) to increase channel capacity of the lower American River in order to safely pass the objective releases of 130,000, 145,000, and 180,000 cfs (Table 2.6). Necessary channel and

TABLE 2.5 Potential Levee Failure

Levee Reach	Remaining Freeboard at Failure[a] (feet)	Stage or Flows[b]	Return Period (years)
Reclamation District 1000			
Sacramento River (left bank) NCC to NEMDC	3	—[c]	—
NCC (north and south levees)	2[d]	40.0 feet	200
NEMDC (west levee)	1.5[d]	35.4 feet	71
American River Levee System			
North (right) bank, Sacramento River to river mile 5.2	3	180,000+ cfs	85+
North (right) bank, upstream from river mile 5.2	4	140,000 cfs	71
South (left) bank, Sacramento River to river mile 5.2	5	140,000 cfs	71
South (left) bank, river mile 5.2 to river mile 7.8	5	145,000 cfs	73
South (left) bank, upstream from river mile 7.8	4	200,000 cfs[e]	94
Dry and Arcade creeks, and east levee of the NEMDC	3	—[f]	—
Sacramento River east (left) bank from the American River to Freeport	3	—[d]	—
Sacramento River west (right) bank from the Sacramento Bypass to Riverview	3	—[d]	—
Yolo Bypass and Tributary levees	3	—[d]	—
Sacramento River west (right) bank from the NCC to the Sacramento Bypass	—[g]	—	—

NOTE: For flood damage estimates only. Actual levee failures may occur at higher or lower flows and stages.

[a]Assumptions: (a) levee rehabilitation as part of the Sacramento River Flood Control and Sacramento River Bank Protection Projects in Sacramento area has been completed, and (b) the remaining sediment in Fremont Weir has been removed.

[b]Unless otherwise noted, flows are at Fair Oaks gage.

[c]Not applicable due to failure at other locations reducing threat.

[d]Freeboard encroached condition chosen based on February 1986 flood conditions.

[e]Nondamaging flow is approximately 145,000 cfs.

[f]Levee failure is not the cause of flood damage on Dry Creek.

[g]For evaluation of flood damages, zero remaining freeboard was selected to be consistent with FEMA's approach to establishing failures.

SOURCE: USACE, Sacramento District, 1991.

TABLE 2.6 Summary of Levee and Channel Modifications to Increase Channel Capacity of Lower American River

	Objective Release		
	130,000 cfs	145,000 cfs	180,000 cfs
Lower American River (miles)			
Slurry wall	0.7	0.9	4.1
Toe drain	0.6	2.7	7.8
New levee	0.9	1.0	1.0
Levee raising	0.0	2.7	11.4
Riprap on bank	1.5	1.5	1.5
Riprap on levee	5.3	5.3	5.3
Riprap on bank and levee	3.2	3.2	3.2
Yolo Bypass	Extensive levee raising on both sides south of Sacramento Bypass		
Sacramento Weir	Lengthen 500 feet	Lengthen 1,400 feet	Lengthen 3,600 feet
Other	Raise Union Pacific Railroad Relocate American River Parkway Access Road Replace Main Avenue Bridge	Replace Main Avenue Bridge and Norwood Avenue Bridge	Raise H Street bridge; replace El Camino, Howe Avenue, Main Avenue, and Norwood Avenue bridges; replace American River bike trail; replace fencing

SOURCE: USACE, Sacramento District, 1991.

levee modifications included slurry walls, toe drains, new levees, levee raising, bank riprap, levee riprap, and various combinations of these projects.

Subsequent Investigations

The 1991 ARWI generally concluded that the levee system was stable for the original design flow (i.e., objective flow) of 115,000 cfs but needed significant remedial work if flows were to be increased to 130,000 cfs or higher.

A subsequent report prepared by WRC-Environmental and Mitchell Swanson and Associates (1992) reviewed the 1991 ARWI. A major conclusion from this review was that there was very little difference between the hydraulic characteristics of 115,000 cfs and 130,000 cfs and, therefore, that the system was not safe for the design flow of 115,000 cfs.

In addition, a later report by Resource Consultants and Engineers, Inc. (1993) was prepared for the Sacramento District. The scope of this effort involved a technical review of the stability and seepage analysis in the 1991 ARWI. The review and subsequent analysis were accomplished using the existing published data; no additional field investigations, borings, or soil testings were performed for this third report. The Resource Consultants and Engineers, Inc. (1993) report concluded that,

> In general, the results of the Army Corps of Engineers analysis are reasonable given the assumptions listed above. They show that seepage pressures and the potential for piping failures will go up significantly as the flows are increased above the 115,000 cfs level.
>
> However, the analysis lacks adequate detail and site-specific data to conclusively evaluate the relative stability of the entire levee reach at the flow level of 115,000 cfs. Exit gradients at the landward side can be much higher when a thin confined layer of pervious materials exists either within the levee or the foundation. The stratigraphy of the section can be as important as the value of permeability selected. The permeability test data were based on only two tests of remolded samples.

The Resource Consultants and Engineers, Inc. (1993) report also concluded that evaluation of the levee stability analysis in the 1991 ARWI indicated that substantially more information and data are required to evaluate levee stability. The report concluded,

> Because layering in the levee foundations is important in the assessment of levee stability, it is recommended that foundation investigation borings be conducted. Additional levee and foundation configuration should be analyzed for potential seepage and piping problems. Further triaxial shear of soil materials should be carried out to better define the range of conditions within the levees. Stability analysis should be revised after the foundation conditions and the soil strength parameters have been verified.

In general, the Resource Consultants and Engineers, Inc. (1993) investigation emphasized the need for significantly more information and data for purposes of evaluating levee stability.

Recent Work

A new risk and uncertainty methodology is under development by USACE, and that methodology was extended by the Sacramento District for this study. The committee was provided with a series of working papers and calculations concerning the risk and uncertainty analysis being developed by the Sacramento District for evaluating the various flood control alternatives, including levees, under consideration for the American River. The 1994 Alternatives Report

(USACE, Sacramento District, 1994a) presents the results of the flood control alternatives evaluation using the risk and uncertainty analysis procedures developed by USACE. The 1994 Alternatives Report, however, does not provide additional information concerning the actual calculation procedures employed in the risk and uncertainty analysis.

With respect to the analysis of channel capacity to convey the objective releases, USACE will no longer treat stage-discharge and stage-damage functions in a deterministic fashion, but will regard these as stochastic functions and will estimate probability distributions for these functions. Risk and uncertainty are incorporated into the analysis by estimating the probable nonfailure point (PNP) and the probable failure point (PFP). The PNP is defined as the water surface elevation below which it is highly unlikely (probability zero) that the levee would fail; the failure probability jumps to 15 percent when the water surface elevation rises above the PNP. The PFP is defined as the water surface elevation above which it is highly likely (probability one) that the levee would fail. For a water surface elevation just below the PFP, the levee would have an 85 percent chance of failure. The failure probability is assumed to vary linearly from 15 to 85 percent for water-surface elevations between the PNP and the PFP. Representative PNP and PFP elevations for the levees were identified at each index location for the without-project conditions. For each alternative the PNP and PFP would be modified to represent the levee modifications proposed for that alternative (USACE, Sacramento District, 1994). The percentage chance of levee failure is then calculated based on the PNP and PFP elevations. Therefore, the selection of the PNP and PFP is an important step in the risk and uncertainty analysis. It appears that no additional data were available for estimating the PNP and the PFP elevations and that the recommendations of the Resource Consultants and Engineers, Inc. report (1993) concerning insufficient data for determining levee failure are still valid.

These efforts by USACE to incorporate risk analysis procedures into decision-making concerning the adequacy of the levees are to be commended, but it is apparent that existing data is insufficient to permit effective application of risk analysis to this decisionmaking process, especially with respect to the levees and the estimation of the PNP and PFP elevations. More data are required to complete the evaluation of levee stability analysis. Consequently, unless the additional data detailed in the Resource Consultants and Engineers, Inc. (1993) report are developed, it would appear that the PFP and PNP elevations and probabilities are no more reliable than the qualitative engineering judgments in the 1991 ARWI and that use of risk and uncertainty in evaluation of alternatives for the American River flood control project will not necessarily increase the quality of the decisionmaking data base.

Given that application of risk analysis procedures will require additional data in order to quantify the parameters of the probability distributions, it appears that a first step in applying this risk analysis procedure should be to acquire the additional data detailed in the Resource Consultants and Engineers, Inc. (1993) recommendations.

Because the various reports available express uncertainty concerning levee stability, the committee has concerns about the Sacramento District's proposed alternatives for repairing and enlarging the levees to permit conveyance of "objective releases" from Folsom Reservoir larger than 115,000 cfs. Before alternatives involving raising and enlarging the levees to permit conveyance of 130,000, 145,000, or 180,000 cfs are used in the flood damage reduction project, sufficient data concerning levee stability must be available to provide assurance that the repaired or raised levees can contain these higher flows. USACE's use of risk and reliability analysis does not eliminate the need for additional levee stability data.

A recent report by the Sacramento Area Flood Control Agency's Lower American River Task Force also supports the need for additional geotechnical evaluation of federal and nonfederal levees with respect to seepage and stability at objective releases greater than 115,000 cfs (Lower American River Task Force, 1994).

OTHER TECHNICAL ISSUES:
FLOOD RECORD AND GEOMORPHOLOGY

In reviewing the analysis performed by the Sacramento District in support of the 1991 ARWI and the criticism of this analysis, the committee identified two critical issues that had not been given adequate consideration: the apparent nonrandomness of the American River flood series and geomorphic issues that affect the long-term stability of the lower American River channel.

American River Flood Record

A critical issue in the management of Sacramento's flood risk is the fact that a high percentage of the largest flows in the unregulated American River flood series at Fair Oaks occurred after the design of Folsom Dam in 1945 (Figure 2.2). For example, in the series of unregulated maximum daily flows extending from 1907 through 1986, the top six flows occurred after 1950. This has meant that the apparent magnitude of the American River flood threat is now substantially greater than what was apparent when Folsom Dam was designed.

Figure 2.3 illustrates the effect of the apparent increase in flood magnitudes on the estimated quantiles for the 3-day rain floods. Shown are the Bulletin 17B frequency curves for the periods from 1905 to 1949, 1950 to 1986, and 1905 to 1986. Also shown are the one-sided upper 95 percent confidence interval for the period from 1905 to 1949 and the one-sided lower 95 percent confidence limit for the period from 1950 to 1986. Note that the estimated 200-year 3-day flood

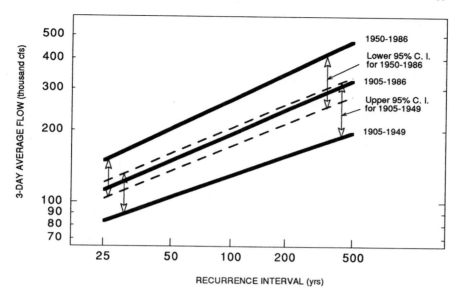

FIGURE 2.3 Split-record quantile estimates for 3-day deregulated rain floods, American River at Fair Oaks, estimated by Bulletin 17B. Confidence intervals shown are one-sided.

volume for the period from 1950 to 1986 is greater than the 500-year 3-day flood volume for the full period.

How should we interpret this apparent increase in flood magnitudes? There are several potential interpretations, with differing implications. These include ordinary random variability, nonrandom climatic variability, changes in the watershed, and nonrandomness due to systematic measurement errors.

It is possible that the concentration of high flows in the unregulated Fair Oaks data is simply the result of random variability. For example, there is about a 3 percent chance that the top six values in a random series of 80 occur in the first or last 40 values. Such an occurrence has a low probability, but it is certainly possible.

Although it is common practice to assume that annual flood series are random, it is widely understood that climate itself is nonrandom, even at time scales of a few years. (Nonrandomness is defined here to include nonstationarity and correlation.) For example, it has been demonstrated that the quasi-periodic occurrence of the El Nino/Southern Oscillation in the North Pacific Ocean correlates significantly with precipitation in the western United States (Swetnam and Betancourt, 1990; Redmond and Kock, 1991; Woolhiser et al., 1993). However, similar relationships have not been found for floods. And although a few investigators have criticized the operational assumption that floods are random (Knox, 1983; Hirschboeck, 1988), little has been offered in the way of alternatives to conventional analysis. One exception is Booy and Lye (1989), which discusses a simple method for accounting for the increased uncertainty in flood risk assessment due to apparent nonrandom climatic variability.

The committee was not able to find any literature relating the apparent increase in large floods on the American River to nonrandom climatic variability. However, there have been relevant studies on seasonal streamflow in the American River and nearby rivers. Roos (1987) documented a reduction in the April–July fractional flow from snowmelt in northern Sierra Nevada streams and suggested that the cause was slight shifts in seasonal precipitation and broad-scale temperature patterns. (The fractional flow for a given period is the ratio of the runoff for that period to the annual runoff.) Pupacko (1993) analyzed streamflow data from the North Fork of the American River and the East and West Fork of the Carson River and documented a trend of increasing and more variable winter streamflow beginning in the 1960s. He attributed this trend to small increases in temperature, which increase the rain-to-snow ratio at lower altitudes and cause the snowpack to melt earlier in the season at higher altitudes.

Aguado et al. (1992) attributed much of the decline in April–July fractional flow to an increase in fractional streamflow earlier in the winter, caused by an increase in precipitation in late fall and early winter, and especially in November. The increase in November precipitation is important, since early season precipitation is more likely to run off immediately rather than be stored in snowpack. Aguado et al. (1992) interpreted the decline in spring-summer fractional flow to be a natural climate fluctuation, rather than a greenhouse warming effect.

How does this apparent shift in seasonal flow relate to flood flows? A climate change that caused flood-producing storms to occur earlier would result in greater flood magnitudes. The most significant floods on the American River occur in the rainy season, which begins in late fall and early winter. Storms that occur later in the winter are likely to have more precipitation in the form of snow, which does not generally contribute to storm-induced flooding. However, the timing of rainfall floods on the American River does not appear to have shifted during the period of record at the Fair Oaks gage. Apparently, the increase in flood magnitudes on the American River is not directly caused by the apparent shift in seasonal flows. It is possible, however, that both are caused by the same climate change.

Tree-ring analysis provides evidence that the climate in the region is subject to nonrandom variability. Using dendroclimatic methods, Earle (1991) reconstructed streamflows since A.D. 1560 for the Sacramento, Feather, Yuba, and American rivers. These reconstructions indicate that there have been a number of periods of prolonged high and low seasonal flows during the past 420 years. Furthermore, the period from 1917 to 1950 is the most extreme dry period in the reconstructed record. It is possible that the end of the dry period, the change in the timing of seasonal flows, and the increase in flood magnitudes are all caused by the same nonrandom behavior of the climate system.

Nonrandomness in the unregulated Fair Oaks series could also result from changes in land use and land management practices in the watershed. The primary land use in the watershed is forestry. The committee did not discover any documentation of changes in forest practices over time. While it is possible that there have been some changes in land use practices, it is unlikely that they alone would have caused the observed increase in large floods. Nevertheless, future land use may change in the American River watershed. Changes from the current land use, which is primarily forest, would most likely increase flood magnitudes.

Systematic errors in flow monitoring and in the adjustment of the gaged flows for the effects of regulation by Folsom Reservoir are another potential cause of the apparent nonrandomness of the Fair Oaks unregulated series. The gage has been located at several sites and was non recording prior to 1930. As discussed in the next section, the river channel throughout the Fair Oaks area has degraded significantly through time. The stage-discharge relation used to monitor the flows has been re-determined frequently to reflect the changing hydraulic conditions. However, as the main channel has degraded, the flow at which over-bank flow begins has gradually increased. Because over-bank flows are subject to more measurement uncertainty than within-banks flows, it is possible that there may be more uncertainty in the measurements of very large discharges earlier in the period of record than in more recent measurements. In addition, the closure of Folsom Dam has facilitated flow measurements by reducing peak flows and providing extended periods of steady flow, and thus may have resulted in recent flood flow measurements having less uncertainty than those early in the record. Nonetheless, there is no evidence that flow measurement uncertainty has contributed to the apparent nonrandomness of the Fair Oaks flood record. It also is possible that there might be systematic error in the post-Folsom unregulated flood record as a result of the method used to adjust the gaged flows for effects of regulation by Folsom Dam. Based on comparisons of the adjusted and unadjusted flows with Folsom reservoirs contents, however, this also appears highly unlikely.

If it is true that the apparent increase in the magnitude of large floods in the American River is the result of nonrandom climatic variability, what are the implications for estimating the probability of large floods? Clearly, one implication would be that probability estimates are more uncertain than would be expected from conventional analysis. Is the flood record since 1950 more representative of the immediate future than the prior record? If so, use of the full flood record would greatly underestimate the flood risk. There also is the possibility that global climate change will increase flood peaks on the American River, by increasing either the amount of precipitation or the proportion of precipitation that falls as rain in storm events (Lettenmaier and Gan, 1990; Roos, 1994). In summary, there is significant uncertainty about the risk of flooding in the American River, much greater uncertainty than would normally be assumed on the basis of the long streamflow record. It would be prudent to explore the economic and safety implications of estimating flood risk in the American River basin using just the second half of the American River flood record, from 1950 to the present.

Geomorphology of the Lower American River

The evaluation of possible flood management alternatives needs to consider pertinent aspects of the local geomorphology. Two geomorphological features are particularly important to flood hazard management in the American River basin: (1) stability of channel banks and levees and (2) potential ongoing channel enlargement and increases in conveyance. Insights about these issues can be gained by examining geomorphic stability and channel erosion in the lower American River, as evidenced by data from U.S. Geological Survey stream-flow measurements at the Fair Oaks gage.

Evaluations of potential temporal changes in flood conveyance in the lower American River must consider channel stability, which in turn is dependent on channel morphology and stratigraphy. Since both morphology and stratigraphy of the lower American are largely the result of extreme and persistent channel changes induced by human activities, analyses of channel stability should begin with an understanding of the nature of historical sedimentation and subsequent channel adjustments.

Historical Channel Changes

The channel geomorphic history over the last 130 years is one of great change. Mining sediment, dams, and levees caused perturbations to which the fluvial system is still adjusting. Channel stability is related to these extensive but undocumented changes due to both nineteenth century aggradation and engineering works. Yet an analysis of the historical record of channel changes that could reveal instabilities has not been conducted. For example, two apparent nineteenth century channel diversions near downtown Sacramento, including a meander-bend cutoff near Sutter's Landing in 1862 and a northward diversion of the channel at its confluence with the Sacramento River (Bischofberger, 1975; Dillinger, 1991), are not mentioned in the geotechnical literature. These changes could represent channel shortening and steepening in critical reaches below Howe Avenue.

Mining Sediment

Hydraulic mining, invented in northern California in 1853, is a method of resource extraction that uses pressurized water to move large volumes of sediment (Paul, 1947; James, 1994). As hydraulic gold-mining came into widespread use in California, much sediment was delivered to main channels of northern Sierra rivers and caused channel aggradation (Mendell, 1881). The lower American River aggraded substantially during the primary hydraulic mining period (1861 to 1884) and later degraded as sediment inflows decreased and channel deposits were transported downstream (Gilbert, 1917). Estimates of mining sediment stored along upper American River channels near the turn of the century were 20 to 25 million cubic yards in the North Fork, 10 to 15 million cubic yards in the Middle Fork, and none in the South Fork (Manson, 1882; Gilbert, 1917). Licensed hydraulic mining continued to produce sediment from 1893 through at least the 1930s. Few of the sediment detention structures required for licensing remain, so most of this sediment was delivered downstream until the North Fork Dam was built in 1939. Little mining sediment remains in the mountains other than sediment stored behind North Fork Dam and Folsom Dam, although a low gravel terrace remains on the North Fork above Lake Clementine (behind North Fork Dam).

In the lower American River, mining sediment deposits were estimated to have varied between 5 and 30 feet in depth across almost 10 square miles (Mendell, 1881; Manson, 1882), and mining sediment still dominates the active sediment. Field visits in 1994 located much historical sediment stored along the lower American River. On the basis of the mineralogy of the pebbles, it was determined that much of this sediment was produced by hydraulic mining (James, 1991b). A left-bank historical terrace 4 m high of erodible unconsolidated sand and gravel at river mile 21 is representative of historical deposits in the lower American River from river mile 15 to 22 (Photos 2.1 and 2.2). The high terrace of historical sediment on the left bank at Watt Avenue (Photo 2.3), extends laterally beneath the levees on both sides of the river and downstream below H Street through an area of critical bank erosion potential.

Channel instability may arise from the morphologic changes induced by historical aggradation. Erosion of historical sediment could be relevant to conveyance in two ways: (1) eroded sediment may fill channels or produce bedforms and other roughness elements during floods, thereby reducing conveyance and raising flow stages, or (2) increased channel capacities could improve flood conveyance. In addition, many levees are built on stratified mining sediment with high lateral hydraulic conductivities. Seepage beneath levees was observed in 1986 and is a substantial problem (RCE, 1993).

Geomorphic mapping is needed to identify where banks and levee foundations are composed of relatively erodible and permeable historic sediment. Two recent reports classified bank stratigraphy along the lower American as Pleistocene or Recent (Holocene) without distinguishing between prehistoric Holocene and historical sediment (WET, 1991; RCE, 1993). No mapping of long-term historical channel changes or field descriptions of present historical deposits has been attempted in the lower river. Nor has the condition of the pre-mining channel been considered, other than base level changes.

PHOTO 2.1 Hydraulic mining was common in northern California in the late 1800s and delivered significant amounts of sediment into Sierra rivers. These historical deposits are visible, such as this terrace near river mile 21. A bike trail runs on top of the terrace. (Allan James, University of South Carolina.)

PHOTO 2.2 Hydraulic mining sediment deposits often consist of erodible, unconsolidated sand and gravel. At river mile 21, about six feet of historical sediment cap about 3 feet of older sediment. (Allan James, University of South Carolina.)

PHOTO 2.3 The potential for channel instability is increased in areas with hydraulic mining deposits, such as this site along the left bank of the American River near Watt Avenue. The terrace surface extends upstream and down-stream, as well as beneath the levee. (Allan James, University of South Carolina.)

Bank and Lateral Stability

As pointed out in the "Geotechnical Analysis" section above, the 1991 ARWI states that banks and levees were structurally stable at flows up to 115,000 cfs, but would fail due to seepage or overtopping at higher flows. The 1991 ARWI was based largely on a geotechnical perspective, neglecting geomorphic processes. Three recent reports have introduced the geomorphic perspective (WET, 1991; WRC-Environmental and Swanson, 1992; RCE, 1993). On the basis of historical aerial photographs and field evidence, consultants for SAFCA (WRC-Swanson, 1992) concluded that bank erosion potential is high, and that sustained bank erosion since 1955 can be attributed to Folsom Dam closure and levee construction.

Consultants for the District identified lateral instability and seepage failures as serious concerns, although the District does not believe that the bank erosion problem goes beyond what can be treated by standard maintenance practices (Sadoff, 1992). Bank stability was evaluated based on stream power, which was highest in steep upper reaches below Folsom, where channels were presumed stable because of resistant strata in the bed and right banks (RCE, 1993). However, extensive deposits of historical sediment on the left bank of these reaches could be prone to erosion. In the lower reaches, stream powers were high between river mile (RM) 5 and RM 6, corroborating other findings that the bends below Howe Avenue are vulnerable to bank erosion. Comparisons of aerial photographs from 1968 and 1986 indicated that channel migration rates at five critical sites (RM 12.5 to 20.1) averaged 4.8 ft/yr and ranged between 1.1 and 8.0 ft/yr (WET, 1991). Migration rates as high as 13.9 ft/yr at other sites were not deemed critical because of the channel distance from a 50-foot buffer around the toes of levee slopes. These migration rates do not include substantial channel changes from the 1965 flood, which caused an avulsion near river mile 15.

Agreement on the potential for lateral channel migration is important not only to bank stability, but also to channel enlargement. Lateral planation in meandering alluvial channels can maintain a natural equilibrium system, but with the down-valley sediment supply cut off by dams, eroded bank material may not be entirely replaced and erosion could result in net channel enlargement over time.

Channel Lowering and Enlargement

Questions relevant to channel stability and potential changes in conveyance in the lower American River include the degree and timing of aggradation and degradation, whether channels have returned to presettlement base levels, and whether channel enlargement continues. Dam closures are often associated with channel erosion downstream (Williams and Wolman, 1984), although responses to dams may be complex and may include periods of local aggradation. For example, closure of Oroville Dam in 1968 caused complex channel changes downstream on the Feather River at least through 1975 (Porterfield et al., 1978). It has also been argued that the lower American River has been degrading in recent decades, encouraged by the closure of Folsom Dam and levee construction in the 1950s (WRC-Environmental and Swanson, 1992), although little evidence has been cited.

Vertical Incision

Vertical changes on the lower American River have been the subject of several investigations. Gilbert's (1917) time series of Sacramento River bed elevations just below the American River confluence showed 10 feet of bed aggradation from 1855 to 1890, and about 8 feet of degradation by 1914. These responses to hydraulic mining sediment indicate that the lower American River also must have experienced substantial channel bed aggradation and degradation. Recent studies of historical incision, based primarily on California Debris Commission (CDC, 1907) and subsequent topographic maps (1955 and 1987), identify 10 to 20 feet of degradation in the lower river from 1906 to 1986 and conclude that thalweg incision is ongoing at some locations (WET, 1991; WRC-

Environmental and Swanson, 1992; RCE, 1993). Ten channel cross-sections, resurveyed between 1987 and 1993, showed no systematic change (RCE, 1993), but these surveys were not separated by any major flood events. At some sites the channel bed rests on resistant premining strata, and removal of historical sediment from the bed is complete at these sites (RCE, 1993). Incision of resistant Pleistocene strata can result in sustained channel degradation, however, as on the nearby Bear River in response to a 1955 flood (James, 1991a).

Thalweg profiles indicate that most channel degradation between RM 6 and RM 11 was complete by 1955, but that considerable incision occurred between 1955 and 1987 from RM 6 to the mouth and between RM 11 and RM 14 (RCE, 1993). Channel incision of about 20 feet and considerable channel enlargement had occurred in the lower American River by 1960 (Olmsted and Davis, 1961). Changes in thalweg profiles on 1957 and 1987 maps indicate an average of about 18 feet of incision between RM 2 and 3 (WET, 1991).

Bed stability was modeled using USACE design 100-year hydrographs and the Parker bedload transport equation based on the median bed material size (D_{50}) and Shields entrainment criteria (RCE, 1993). Most simulated channel beds experienced no scour, and maximum bed elevation change under the worst scenario was less than 2 feet (at RM 7). On the basis of the model, channel beds throughout the lower American River should be stable under relatively large and infrequent events.

Channel Enlargement

Vertical incision is only one form of channel erosion, and vertical stability would not preclude channel enlargement by erosion of sediment stored along channel margins. It is common in aggraded systems for channels to respond initially to decreased sediment loads by incising vertically, and later to widen out; particularly when channel top widths are confined by levees or terraces. For example, it has been shown experimentally that knickpoint retreat is often followed by lateral migration and bank erosion (Schumm, 1973; Schumm et al., 1987).

Following vertical regrading of the lower American River channel profile, a period of channel enlargement by bank and berm erosion and lateral migration cannot be ruled out. In fact, due to surplus energy from decreased sediment loads and decreased channel capacities from levees and historical deposits, and due to observed channel erosion and lack of sediment replacement from above Folsom Dam, ongoing net channel erosion could be expected for the lower American River. In spite of these reasons to suspect channel enlargement and the ramifications to channel conveyance and environmental concerns along the parkway, evidence of channel change in the lower American River has not been adequately studied.

Channel Changes at Streamflow Gages

The nature of channel erosion since the closure of Folsom Dam has been examined using topographic maps and aerial photos with limited temporal and spatial resolutions (WRC-Environmental and Swanson, 1992; RCE, 1993). To enhance the channel-change data base, the committee examined high-resolution U.S. Geological Survey cross-section measurements at the Fair Oaks gage. These analyses are based on only a few sites associated with various locations of Fair Oaks gages and soundings, so caution should be exercised before extrapolating results up- or down-stream.

Channel changes are demonstrated by channel cross-sections and stage-discharge regression analysis. Data were derived from stream-flow measurement records (USGS archives). Cross-section plots were derived from depth soundings at three locations (Figure 2.4): the old Fair Oaks Bridge (1913 to 1950), a cable about 300 feet below the bridge (1930 to 1957), and a cable 2.2 miles upstream below Hazel Street (1958 to 1994). All sections are from bridges or cables to control the longitudinal position. Numerous plots reproduced sections during stable periods indicating high accuracy of the procedure. For the sake of brevity, only five cross-sections at one site are presented here.

Channel morphological changes are rarely related to changes in flood stages

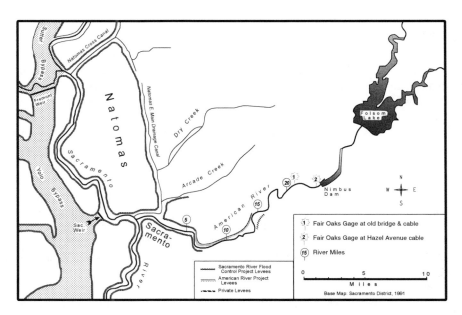

FIGURE 2.4 Locations of gages and levees on the lower American River. SOURCE: A. James, University of South Carolina (adapted from USACE, 1991).

TABLE 2.7 Stage-Discharge Data

Location	Total N	Model N	Model Years	Q Range	R^2
Bridge	528	497	1905 to 1958	$500 < Q < 20,000$	0.85
			Stage $= 67.5 + 7.18 \cdot 10^{-4}\, Q - 2.00 - 10^{-8}\, Q^2 + 2.30 - 10^{-13}\, Q^3$		
Hazel Street	454	413	1958 to 1994	$Q < 15,000$	0.74
			Stage $= 76.1 + 1.52 \cdot 10^3\, Q - 1.66 \;\; 10^7\, Q^2 \;\; 7.05 \;\; 10^{-12}\, Q^3$		

in a simple manner. For example, main channel deepening may not result in lower stages of overbank floods if meander-belt flows develop greater turbulence at channel crossings (Ervine et al., 1993). Thus, an independent analysis of stage-discharge relationships was conducted to evaluate temporal changes in stage at the two gage sites: the old Fair Oaks Bridge and Hazel Street sites. Stage integrates morphologic and hydraulic factors, providing an indicator of flow conveyance. Stage data represent gage readings at the time of discharge measurements (not rating curves), corrected for gage datum changes.

Flow stage is strongly related to discharge, so stage was statistically regressed on discharge to control for these effects. A third-order polynomial provided the best-fit model at both sites. Extreme discharge events were eliminated from regressions (Q-Range, Table 2.7) to emphasize changes within the inner channel rather than overbank characteristics that can be dominated by roughness elements. The regressions provide an objective estimate of the stage of a given discharge. Plots of residuals (errors in the predicted stage) against time reveal temporal changes in stages of flows up to moderate magnitude floods. These methods and some limitations to their morphologic interpretation (e.g., changes in roughness and energy gradient) are explained elsewhere (Knighton, 1974; James, 1991a).

Fair Oaks Gage near Old Bridge

Cross-section plots (1913 to 1950) at the Fair Oaks bridge indicate channel bed scour and fill with net thalweg erosion of about 8 feet (Figure 2.5). Channel morphology is controlled by bridge piers and the right-bank bluff. A deep left-bank fill narrowed the channel by about 20 feet toward the end of the period suggesting that constriction by the bridge is not the dominant reason for erosion. A cable was installed about 300 feet below the bridge in 1930, where cross-section plots indicate about 5 feet of thalweg erosion from 1944 to 1952 followed by about 2 feet of fill by 1957 when the cable was moved. Deepening and narrowing of cross-sections at this site suggest that erosion at the bridge extended through the reach. Channel deepening and narrowing at this bridge site follow

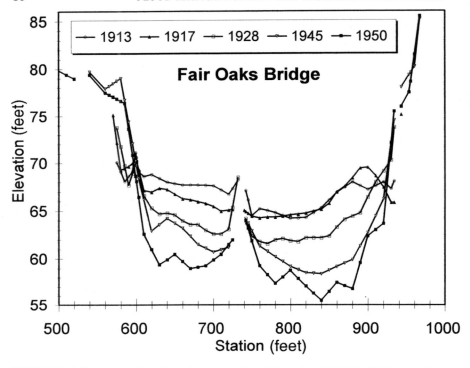

FIGURE 2.5 Representative channel cross-section plots at the Fair Oaks Bridge showing about 8 feet of thalweg degradation between 1913 and 1950. Data gaps indicate bridge piers. The view is downstream.

the general response observed elsewhere where channels are incised through hydraulic mining sediment (James, 1991a).

Stage-discharge relationships at the bridge site indicate a systematic grouping by period with occasional changes in flow stages (Figure 2.6). Temporal patterns of flood stage changes are illustrated by a time series plot of regression residuals (Figure 2.7). Flow stages at the old Fair Oaks gage rose slightly from 1905 to 1912, lowered about 2 feet by 1920, rose about 2.5 feet in the late 1930s, and dropped about 3.5 feet by 1950 to about 1.5 foot below the mean for the period. The rapid incision during the 1940s may represent a response to decreased sediment yields following the closure of North Fork Dam in 1939.

Fair Oaks Gage at Hazel Avenue

In 1957 the gage and cable were moved 2.2 miles upstream to the present Hazel Street site below Nimbus Dam. From 1958 to 1994 the channel at this

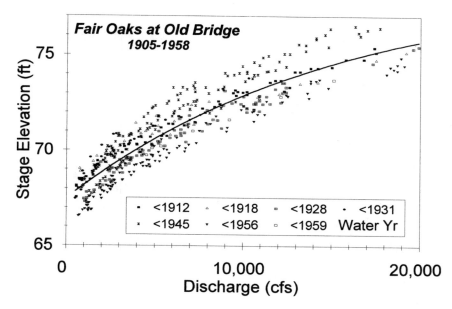

FIGURE 2.6 Stage-discharge relationship from the Fair Oaks gage at the bridge and early cable site. Several distinct periods of high and low stages can be identified.

location experienced episodes of thalweg deepening and bar deposition followed by stable periods lasting several years, and about 9 feet of net thalweg degradation. The 1965 flood scoured the thalweg about 10 feet, but the channel partly refilled from 1965 to 1973 and was colonized by willows. From 1973 to 1986, the channel bed was stable, but the 1986 flood lowered the thalweg about 3 feet and widened the channel considerably.

Analysis of flood stages at the Hazel Street site indicates two periods of relative stability from 1958 to 1967 (Figure 2.8). Stage-discharge regression residuals reveal lowering of flow stages at this site, between the two stable periods (Figure 2.7). The 1965 scour event had no effect on flow stages, presumably due to rapid refilling and increased vegetational roughness of the channel. From 1967 to 1970, however, flow stages rapidly lowered about 2 feet. Sustained incision over the period from 1958 to 1994, during which time flow stages dropped about 2 feet, suggests a long-term tendency for channel degradation and a mobile bed at this site. The close proximity of Nimbus Dam upstream severely limits replacement of eroded bed sediment, resulting in net degradation.

Thalweg incision at the two gage sites was about 8 feet (1913-1950) and 9 feet (1958 to 1994), respectively. Although net stage lowering for the two periods was only about 1.75 feet and 2.5 feet, respectively, large rapid fluctuations characterize these changes. This evidence of rapid erosion at gages lends

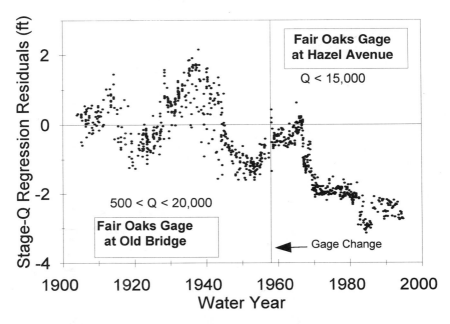

FIGURE 2.7 Stage-discharge regression residuals for the Fair Oaks gage. Left side is from Fair Oaks bridge site (see Figure 2.6) and shows two periods of low stages and two of high stages interpreted as degradation and aggradation, respectively. Right side is from Hazel Avenue site (see Figure 2.8) and shows a short period of rapid stage lowering interpreted as in response to channel degradation. Joining of the two series is approximate.

credence to a hypothesis of continued channel deepening and enlargement in the upper reaches. If the gage sites are representative of other sections, the conclusion that extreme floods would cause little incision on the lower American River (RCE, 1993) could underestimate the potential for channel down-cutting.

Geomorphic Conclusion

Bank stability is a serious consideration when considering conveyance of high flows in the lower American River. Although the degree of hazard that bank erosion, lateral migration, or bed incision pose to levee stability is contested, all parties appear to agree that a program of channel monitoring and maintenance is necessary. The belief that historical sediment in channels of the Sacramento Valley is now stable is based largely on evidence of elevations derived from topographic maps and numerical simulations of channel bed erosion. Thalweg

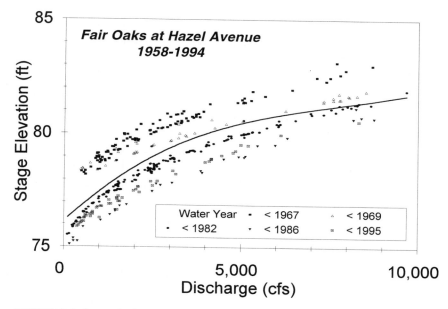

FIGURE 2.8 Stage-discharge relationship from the Fair Oaks gage at the Hazel Avenue cable site. Several distinct periods of high and low stages can be identified.

elevations indicate that base-level adjustments have decelerated, but ongoing vertical adjustments should not be ruled out. Nor would stabilization of long profiles necessarily indicate an end to channel bank erosion, lateral migration, enlargement, or instability.

Evidence from two Fair Oaks gage sites indicates substantial local channel bed scour. From 1913 to 1958, flow stages at the Fair Oaks bridge changed considerably, showing two periods of increasing stages and two of decreasing stages, interpreted as periods of aggradation and degradation, respectively. There was a net lowering of flow stages by almost 2 feet for this period, presumably due to erosion of historical sediment. From 1958 to 1994, flow stages at Hazel Street also lowered about 2 feet. If these sites are representative of the lower river as a whole, further channel incision may be anticipated.

Given historical aggradation, cessation of sediment deliveries since dam construction, and evidence of erosion, the potential for net erosional tendencies in the lower river cannot be rejected. A sediment budget deficit exists in the lower river as dams arrest sediment deliveries from upstream while erosion removes sediment, and this deficit results in net erosion. The hypothesis that channel erosion and enlargement have resulted in increased channel conveyance over the last two decades should be tested further using hydraulic models. Analysis of stage-discharge time series provides empirical support for the hypothesis that channel stages of moderate magnitude floods have lowered by a modest amount at two locations over two different periods, but more information is needed to substantiate these results and extend them to other locations downstream.

Given the critical nature of flood hazards in Sacramento and extensive nineteenth century channel changes, the committee suggests three areas of study regarding the geomorphology of the lower American River: (1) ongoing monitoring of channel changes, (2) historical reconstruction of channel changes, and (3) geomorphic mapping. Recent and ongoing channel changes should be documented and monitored following large flood events by repeating channel cross-section surveys, and by registering aerial photos.

Study of long-term historical changes should include consultation of early historical records to establish presettlement channel conditions that could establish a baseline for changes to the fluvial regime that was presumably in equilibrium with long-term flow conditions. In addition, historical changes should be documented through historical and field methods. For example, CalTrans bridge surveys could be collected and repeated, and California Debris Commission records of twentieth century hydraulic mining sediment production could be tabulated.

Vast tracts of erodible historical sediment stored in the lower river should be studied and mapped. In the upper reaches they are relevant to channel enlargement and sediment production, while in the lower reaches they are relevant to bank and levee stability and seepage. Mapping will reveal spatial patterns and allow more accurate interpolation between geotechnical sample points. As pointed out above, implementation of risk and uncertainty analysis in the lower American River will require appraisals of channel and levee stability (USACE, Sacramento District, 1994a). Assignment of PNP and PFP elevation for levees should be based in part on knowledge of lower American River stratigraphy with an emphasis on the spatial pattern of historical sediment and former channels.

3

Environmental Issues

The American River basin possesses significant environmental values. The upper American River, known for its steep gradient, free-flowing white water, and relatively natural plant communities, was listed under the state's Wild and Scenic River classification system in 1978. In 1981, a 23-mile stretch of the lower American River was designated under the National Wild and Scenic River System as a sport fishery and a critical recreational resource. Sacramento owes its nickname, "River City," its identity, its sense of history, and its regional character to the American River, which flows through the city's center, and to the Sacramento River, which joins near the city's historic district.

The 1991 American River Watershed Investigation (ARWI) raised a number of contentious environmental issues. A key concern was whether the description of the environmental impacts of the various proposed project alternatives was adequate. This chapter reviews how the 1991 ARWI and the associated environmental impact statement/environmental impact report (EIS/EIR), referred to here as the 1991 ARWI report (USACE, Sacramento District, 1991), considered environmental impacts. It pays particular attention to the assessment of environmental impacts associated with the proposed Auburn dry dam alternative and associated inundation impacts in the American River canyon, but also addresses a range of other environmental considerations. In addition, the chapter discusses whether the planning process is adequate to produce plans that consider a full array of feasible alternatives that integrate social, environmental, and flood risk reduction factors. (Chapter 6 addresses planning issues in more detail.)

TREATMENT OF ENVIRONMENTAL ISSUES
IN THE ARWI REPORT

Overview of NEPA and CEQA

The 1991 ARWI was prepared to meet the requirements of both the National Environmental Policy Act (NEPA) of 1969 and the California Environmental Quality Act (CEQA). NEPA Section 102 requires an environmental impact statement for "major federal actions significantly affecting the quality of the human environment." The Council on Environmental Quality (CEQ) oversees the implementation of the NEPA act. Guidelines prepared by CEQ and federal agencies and a large number of legal cases have established that the major legal issues associated with the EIS process are determining when one must be done (i.e., determining what is a major federal action) and determining the adequacy of a prepared statement with regard to the accuracy of description, the identification and quantification of probable environmental impacts, and the exploration of reasonable alternatives. CEQA passed a year later and, although based on NEPA, had some different provisions. It is implemented through a state agency, the Office of Planning and Research, and has its own guidelines for preparation of environmental impact reports. Both CEQA and NEPA guidelines urge the issuance of joint reports that satisfy both state and federal law (Heyman, 1974; Remy et al., 1994).

Like the federal act, CEQA was conceived primarily as a means to document and consider the environmental implications of actions. Unlike NEPA, CEQA is not merely a "procedural" statute but contains substantive provisions that agencies are to comply with. Under NEPA the federal government is required only to give "appropriate consideration" to environmental values and presumably can take actions causing environmental damage even if feasible and effective mitigation measures could easily be implemented. In contrast, CEQA requires agencies to implement feasible mitigation measures and alternatives identified in the environmental impact reports. In addition to this important distinction, CEQA places an emphasis on any growth-inducing impacts associated with proposed actions and the potential of population growth to stress existing community service facilities (Secretary for Resources, California, 1973; Remy, 1994).

Both NEPA and CEQA provide guidance on the development of alternatives in the EIS and EIR. CEQ guidelines state (Council on Environmental Quality, 1973):

> A rigorous exploration and objective evaluation of the environmental impacts of all reasonable alternative actions, particularly those that might avoid some or all of the adverse environmental effects, is essential. Sufficient analysis of such alternatives and their environmental costs and impact on the environment should accompany the proposed action through the agency review process in order not to foreclose prematurely options which might have less detrimental effects.

The American River is a significant recreational resource. The lower river is an important sport fishery, while the upper river is a steep, free-flowing river valued by boaters. (Rutherford H. Platt, University of Massachusetts, Amherst.)

Examples of such alternatives include the alternative of taking no action, or of postponing action pending further study; alternatives requiring actions of a significantly different nature which would provide similar benefits with different environmental impacts (e.g., nonstructural alternatives to flood control programs, or mass transit alternatives to highway construction).

CEQA's guidance states, "Attention should be paid to alternatives capable of substantially reducing or eliminating any environmentally adverse impacts, even if these alternatives substantially impede the attainment of the project objectives, and are more costly" (Secretary for Resources, California, 1973).

Environmental Information Deficiencies in the 1991 ARWI

There were several areas where the lack of scientifically based descriptions of environmental impacts prevented the 1991 ARWI from serving as an adequate planning document to assess the impacts of proposed projects. The most significant information deficiencies were in the assessment of (1) potential impacts of periodic inundations from a dry dam on the plant communities located in the upper American River canyon, (2) potential impacts of inundation on canyon soils and geologic stability, and (3) potential impacts on ecosystems and regions. Information gathered since the issuance of the 1991 ARWI is still incomplete, so it is not possible to determine or quantify impacts from inundation by a dry dam in the American River canyon. An adequate EIS/EIR would have described the situation in these terms and made recommendations for a research plan.

NEPA and CEQA guidelines call for impact assessments to consider regional and ecological contexts, but readers of the 1991 ARWI report cannot discern the regional significance of the fisheries, plant communities, wildlife, or the landscape of the upper American River canyon. It is also unclear whether potential ecosystem shifts might result from the cumulative effects of inundation, unstable geology, and plant regeneration problems. The regional significance of such an ecosystem shift was not explored. For example, if the wild trout fishery were affected in the Middle Fork, how would this affect the status of other fisheries in the region or state? Environmental impact assessments must not only identify the probable effects of an action, but also estimate the magnitude and evaluate the importance of these effects. Numerous reports from the early 1970s observed that the assessment process suffered from lack of a systematic means of reporting impact significance and provided procedures to remedy the problem (Leopold et al., 1971; Warner et al., 1974; Dickert and Domeny, 1974). Because of the scale of the controversy associated with building a dam on a scenic river, uncertainties about the significance of impacts will only hinder decisionmaking and efforts to gain public consensus. The measures that have been taken to improve the 1991 ARWI should provide significant benefit. The 1994 Alternatives Report (USACE, Sacramento District, 1994a) contains a more thorough consideration of options to reduce flood damage, but this preliminary document does not provide analysis of the environmental impacts or gains associated with the different alternatives. A supplemental EIS/EIR was not available in time for this committee's review, so it cannot comment on the final planning document.

Project Alternatives Assessment in the 1991 ARWI

To meet the requirements of NEPA, and CEQA in particular, the 1991 ARWI reports should have provided substantial environmental analysis of the different alternatives that could be used to increase conveyance in the lower American River. Environmental restoration project components and geomorphological considerations involving sediment transport and deposition in the lower American and Sacramento rivers, weirs, and bypasses should have been integrated into project alternative scenarios. If geomorphological factors were judged to have no influence on managing or increasing conveyance capabilities in the lower river systems or bypass, then these conclusions should have been substantiated.

Following the release of the 1991 ARWI, a number of significant changes occurred that enabled the 1994 Alternatives Report to present a fuller array of project alternatives. The 1991 ARWI focused strongly on reservoir related options, while the 1994 report was oriented more toward river flood conveyance. Coordination with upstream hydroelectric reservoirs has been arranged to improve Folsom flood operations. The local project sponsor has hired consultants in geomorphology to integrate the rebuilding of lower American River levees with riparian habitat restoration. Before and during the preparation of the 1994

Alternatives Report, the Lower American River Task Force, assembled by the local sponsor, met with the goal of actively soliciting the input of any public and agency stakeholders in the lower American River on environmental and public safety needs (SAFCA, 1994b). A more extensive risk and uncertainty analysis was conducted, which should help Sacramento officials better select risk reduction alternatives that balance public safety, financial, and environmental costs.

During the period between issuance of the two reports, local initiatives were begun in cooperation with the Sacramento District to restore wetlands in the Yolo Bypass. (The first Yolo Bypass wetland restoration projects date back to 1990.) The interagency Yolo Basin Working Group and the District began an assessment of how to integrate both flood protection and environmental restoration objectives in the Yolo Bypass. By 1994, interagency agreements had been approved for the multi-objective management of the Bypass for endangered species protection, wetland and wildlife habitat restoration, and flood control. In 1994 the California Reclamation Board passed Resolution 94-3, requesting that USACE initiate a new reconnaissance study of the Sacramento River Flood Control Project for the purpose of cooperating with other federal and state agencies and public interests leading to a "comprehensive multi-objective river corridor management plan." The local sponsor has realized that the public interest in the environmental values of the upper American River canyons must be an integral component in any flood damage reduction plan and toward this end has reordered its project priorities.

The USACE goal is to design projects and conduct analyses of these designs. Land use planning traditionally has been a local responsibility and this division between the two levels of government often results in the omission of consideration of how the two elements—project design and land use—are interrelated. USACE environmental assessments do not generally deal with the secondary impacts of flood damage reduction alternatives, even though such impacts are possible and indeed probable. For example, in the American River case, if an Auburn dam were built as multi-purpose facility, it might open up significant suburban development in Sierra foothill counties such as Placer and El Dorado Counties. On the other hand, if a single-purpose flood control dam were built at the Auburn site, it might facilitate development in the Natomas Basin. The state of California's CEQA guidelines do require explicit attention to the relationships between land use and facilities development.

A truly comprehensive EIR/EIS, might explore how development options could increase or reduce the need for an Auburn flood control facility, levee improvements, or other measures. It would identify the land use development options near Sacramento with the highest to lowest flood risks. In other words, the description of flood risk reduction alternatives would include land use alternatives and how they can be used to lessen the risk exposure of population growth centers.

A report with an adequate representation of alternatives would also describe how nonstructural measures could be integrated with other measures to form flood risk reduction alternatives. Nonstructural alternatives alone will not provide adequate flood risk reduction for most areas in the American River floodplain, but they can provide important supplements to levee or dam construction scenarios. Easily placed and removed temporary dams for doorways and windows for central Sacramento structures could have been used during flood events such as 1986, when great uncertainties about levee stability and channel capacity on the American River threatened the area. New construction could require elevation of structures both in older developed areas and newly developing areas. A combined flood warning and flood proofing system could be used regionwide.

Current efforts under way by the city of Sacramento to produce a comprehensive floodplain management plan present an opportunity to integrate federal flood control facilities planning along with the National Flood Insurance Program and local plans. The city plan places new emphasis on floodplain zoning, land use planning, floodproofing, flood warning, and evacuation plans. Presumably, the supplemental EIS/EIR preparation is well timed with these local efforts for producing integrated plans. Chapter 5 addresses these floodplain management issues in greater detail.

Limitations of the Environmental Impact Assessment
Approach to Project Planning

Good project planning should be more than just an exercise in disclosure of the potential consequences of a project. The EIS/EIR process can serve as a dynamic planning tool to facilitate development of project alternatives and community consensus. Public and private stakeholders should be involved as co-participants in the identification of reasonable alternatives, rather than merely reviewing draft reports. The Sacramento Area Flood Control Agency's Lower American River Task Force, organized subsequent to release of the 1991 ARWI to plan levee improvement projects, seems to offer a successful model for improved project planning, although this effort is still in progress and thus it is premature to comment on its ultimate usefulness.

The 1991 ARWI identified a preferred alternative plan and described its impacts. It then identified the manner and extent of mitigation that might take place. The concept of environmental mitigation provides an unsatisfactory framework for this water planning process. Environmental mitigation assumes that environmental factors are considered separately, after hydraulic, hydrologic, and engineering factors. One of the conceptual weaknesses of this planning framework is that the objectives of civil works projects and the objectives of restoration projects are viewed as mutually exclusive, competitive objectives or as trade-offs, rather than as mutually supportive objectives. This lack of integration contributed to the reason the 1991 ARWI report did not adequately address ways for managers concerned with reducing flood damage to find incremental gains in reducing flood risk using different management schemes for levees, weirs, and bypasses as well as reservoir reoperations. The pressures placed on the Fish and Wildlife Service (FWS) to provide quantification for plant mortality for a dry Auburn dam, when FWS staff felt quantification based on existing information was not scientifically defensible, is symptomatic of the mitigation-based emphasis. The quantification was forced too early in the planning process for FWS to be able to develop a mitigation plan to make specific projects possible.

In situations where trade-offs between hydraulic or other engineering concerns and natural resources need to occur, mitigation is an appropriate remedy. Mitigation is only as good as its implementation, however, which creates significant uncertainties. In comments on the 1991 ARWI, the public expressed disappointment in past mitigation performances (e.g., of USACE on its Warm Springs Dam and New Melones Dam). In response to this sentiment, project mitigation strategies that address public distrust should be developed.

If environmental features are not determined to be required for mitigation, they are treated as add-ons, or "enhancements," to a project. The current practice of USACE is not to engage in enhancement projects and to relegate these as "separable" projects to the responsibility of local sponsors (Kiesck, 1994). While federal participation in environmental restoration and enhancement projects is fully authorized, planning practice in 1991 and even in 1995 has not yet reflected these legislative policies.

ASSESSING THE IMPACTS OF A DRY DAM

The probable environmental impacts of an Auburn dry dam to the American River canyon are of central importance in determining the desirability of flood risk reduction alternatives. Two critical areas in which information has been deficient are the impacts of periodic inundation on canyon soils and geologic stability and impacts on plant mortality. The significance of these two issues is that the degree to which canyon hillslope failures or vegetation mortality do or do not damage scenic or ecological values has an important bearing on the public perception of the desirability of a dry dam. Landsliding can degrade scenic values, alter the physical and ecological base for plant and wildlife communities, and degrade fisheries through sedimentation of stream channels. A combination of geological instability and inundation stress on plant communities could alter canyon scenery and ecology.

Many Sacramento area residents assign high value to the aesthetic, environmental, and recreational attributes of the scenic canyons of the North and Middle Forks of the American River. To accommodate this public sensitivity, the Sacramento District and local flood control planners, with public input, proposed a "dry dam"—an innovation at this scale. The dry dam at Auburn was conceived as an environmentally sensitive alternative to a previously planned multipurpose dam. As described in the 1991 ARWI, the dam would have impounded peak flows during periods when upstream runoff exceeded the dam outlet capacity. The ungated outlet would have been sized to allow unrestricted passage of normal streamflow volumes, and flow volumes exceeding outlet design would have been detained only for the length of time required to drain through the outlet. This alternative would presumably decrease impacts to vegetation, wildlife, aesthetics and recreation that normally accompany permanent pool multipurpose dams (USACE, Sacramento District, 1991).

The range of technical issues raised concerning the potential dry dam impacts includes how to predict impacts from unstable geology and soils in inundated zones; how to predict impacts from inundation of nonriparian chaparral, oak woodland, and digger pine communities, which are not covered by inundation literature; how to determine the impacts of indirect, longer-term influences stemming from inundation; how to determine the overall ecosystem response to inundation; and how to collect valid data from existing sites with some similarity in plant communities that have undergone inundation events.

Canyon Slope Stability

There is a legitimate concern over hillslope stability in the upper American River canyon, given the frequency of hillslope failures in the region, the condition of the slopes, and the importance of saturation to slope failures. The extent of impacts cannot be quantified because of a lack of empirical data on depths of anticipated drawdowns and long-term effects of vegetation changes on slope stability. This section evaluates what is known about potential effects on slope stability of deep inundation and rapid drawdowns as might occur behind a dry Auburn dam. Further study on the stability of slopes under current conditions and over extended time scales will be necessary if the dry dam option is pursued.

Physical Mechanisms of Slope Failure

Slope failures are common in the Sierra Nevada and west central California, where single storm events have generated numerous failures (Campbell, 1975; Ellen and Fleming, 1987; DeGraff, 1994). Many conditions commonly associated with landslide susceptibility (Cooke and Doornkamp, 1990) pertain to the Middle Fork canyon. Most landslides in California occur during the winter and spring rainy season, which is also the period likely to coincide with deep inundation.

Inundation and rapid dewatering affect slope stability. Rising-stage failures are not a dominant concern in cohesive materials because high pore pressures are offset by hydrostatic forces of the submerging water (Taylor, 1937; Chandler, 1986). They can, however, be a concern in granular, noncohesive materials (Chandler, 1986), such as steep riverwash materials or unconsolidated roadbeds.

Rapid dewatering, on the other hand, may pose a substantial problem in the canyon. Rapid drawdowns can generate slope failures due to increased effective shear stresses while pore pressures remain high (Chandler, 1986), especially in fine-grained materials (Brunsden, 1979). Excess pore pressure is not supported by grains and can be resisted only by soil cohesion (Taylor, 1937). Experimental results indicate that grain-contact stresses may fall to zero, causing pore-water pressures to locally support the entire stress field (Iverson and LaHusen, 1989).

Some failures occur progressively through the cumulative deterioration of friction elements in the matrix until resisting strength is seriously compromised (Brunsden, 1979; Chowdhury, 1992). For example, clay mineral grains may become locally aligned along a failure plane, which facilitates subsequent failures (Chandler, 1986). Laboratory results indicate that pore-water fluctuations can propagate outward from existing shear zones, leading to expansion of shear zones (Iverson and LaHusen, 1989). These factors imply that deep inundation and rapid drawdowns could decrease shear strengths and increase mass wasting hazards long after the inundation period.

Past Hillslope Failures

Landslides generated by rapid drawdown are common. For example, drawdowns during the failure of the Teton Dam in Idaho in 1976 resulted in the failure of about 3.6 million cubic yards of material from the canyon walls (Schuster and Embree, 1980; Cedergren, 1989). In addition, major landslides can cause damming of valleys (Evans, 1986) and dam failures downstream. The Vaiont slide in Italy, which was triggered in part by buoyant forces due to elevated ground water levels in response to reservoir filling, was catastrophic both in volumetric proportions and loss of life, because it led to failure of the dam below (Cedergren, 1989; James and Kiersch, 1991).

There is little empirical information on drawdowns approaching the rates or magnitudes that might occur behind an ungated Auburn dry dam (USACE, Sacramento District, 1991), although rates could be controlled and damage lessened if controllable gates were employed. Drawdowns in the canyon following the cofferdam breach in 1986 generated numerous landslides and provide the most direct indicator of landslide potential at this location, although no slopes were monitored. The 1986 conditions are regarded as a worst-case scenario for a single, isolated geomorphic event, as drawdown was extremely rapid and was completed within a few hours. An inventory and analysis of landslides in the canyon identified and mapped at least 35 small new slides caused by the 1986 inundation and about 5 slides interpreted as reactivated older slides (USACE, Sacramento District, 1991, Appendix M). In addition, two large ancient failures were recognized as having potential to fail, and it was recommended they be monitored. These slides are the River Mile 22.4 Slide, which had part of its toe removed by the 1986 flood, and the Cherokee Flat Slide. The Sacramento District concluded that there have been several episodes of prehistoric slides and that they make up only a small percentage of the total canyon area, but that it is impossible to determine the magnitude and frequency of sliding in the canyon.

The nature of expected canyon inundation was provided explicitly in the form of depth-duration-frequency curves for the original dry dam (USACE, Sacramento District, 1991), although frequencies should have been increased for events with recurrence intervals of less than 10 years due to derivation from an annual maximum rather than a partial duration flood series (Stedinger et al., 1993). A similar set of curves should be generated for any new planned structure. In the initial dry dam plan, the Sacramento District mapped inundation areas using the depth-frequency-duration curves and concluded that inundation impacts would be insignificant, with an estimated 1,927 acres of vegetation lost to combined inundation and mass wasting (USACE, Sacramento District, 1991). It concluded that following a period of inundation-induced slides, stability would soon be reached: "Most likely, each episode of filling and emptying should cause fewer failures as the unstable portions of the slopes are gradually removed and eventually the canyon walls should stabilize" (USACE, Sacramento District, 1991, Appendix M). An alternative model, however, could be that failures will propagate upslope beyond the inundation upper limit (WRC-Environmental and Swanson, 1992) and that slope stability would not be reestablished as long as there is a substantial amount of colluvium at high gradients.

FWS was critical of the proposed dry dam alternative. On the basis of analysis of aerial photographs in the lower canyon, it concluded that slope failures would be substantial and would have adverse effects on vegetation and habitats in the canyon (FWS, 1991). The California Department of Water Resources (CDWR) expanded on the FWS study by analyzing soil maps and aerial photographs further up the canyon, where soils are coarser grained and more permeable (CDWR, 1991). It found that permeable soils tended to remain stable through the 1986 inundation, while prominent scarps developed in impermeable soils. The study concluded that 35 percent of the inundation area is coarse riverwash not susceptible to failure. Of the remaining area, 50 percent (all permeable soils) would be stable under drawdown rates for the 400-year flood control dams but possibly unstable under rates for the 200-year dam (which were proposed at nearly twice those for the 400-year dam), and 15 percent (all impermeable soils) could be unstable under either drawdown rate. These values have been criticized for underestimating potential failures because (1) the soil permeability model did not account for throughflow or rainfall excess contributions from upslope unsaturated zones, (2) the threshold head differential was overestimated, and (3) pool drawdown rates were underestimated (WRC-Environmental and Swanson, 1992).

The CDWR study modeled slope stability by comparing soil drainage rates to drawdown rates. The soil-water approximations, based on a ground water model, appear valid, but the criterion for interpreting those values has been criticized (WRC-Environmental and Swanson, 1992). A critical head differential was estimated between reservoir stage and soil piezometric head above which slope failures were presumed to occur. This threshold was set at 35 feet based on debris-flow scar lengths (including toeslope sediment accumulation areas) measured from the landslide maps. This 35-foot critical head differential has been criticized as too large for several reasons (WRC-Environmental and Swanson, 1992):

• Failures often begin on upper slopes and propagate down across lower slopes, so slide lengths exceed initial failure planes (WRC-Environmental and Swanson, 1992; DeGraff, 1994).

• Failure plane lengths do not necessarily equal head differences at the onset of sliding because failure planes can extend both above the saturated zone and below the reservoir water surface.

• Subsequent analysis located numerous small slides not recognized by the CDWR study and concluded most slides were less than 30 feet in length (WRC-Environmental and Swanson, 1992).

• Critical head differentials derived from the difference between reservoir stage and soil piezometric head were compared by CDWR (1991) to much slower drawdown rates than those anticipated (WRC-Environmental and Swanson, 1992).

Fugro-McClelland and Leiser (1991) suggested that the CDWR estimates of areas susceptible to mass wasting were excessive because they neglected the stabilizing effects of vegetation roots. These authors subtracted the entire area of fine-grained soils (400 acres) on the premise that those slopes would ultimately fail whether inundated or not, and lowered the remaining area of susceptible coarse-grained soils by 50 percent, from 2,200 to 1,100 acres, on the premise of root stabilization. This assessment was arbitrary and inappropriate because the model did not analyze slope shear strengths, it evaluated slope stability using an empirical relation between head differentials and landslide scar lengths. There is no physical basis for decreasing the CDWR (1991) estimated areas of slope stability to account for vegetation roots.

Landslide scar length distributions, a slope-elevation-frequency graph, drawdown analyses, and a slope failure frequency by elevation curve were presented by WRC-Environmental and Swanson (1992). They concluded that stresses on slopes will be much greater than CDWR (1991) estimates, that drawdown rates greater than 3 ft/hr (3 times the CDWR threshold) will occur on more than 50 percent of the slopes, and that about 80 percent of the inundated slopes (2,300 acres) are "extremely likely to fail" (WRC-Environmental and Swanson, 1992). They called for analyses to establish critical drawdown rates, stable slope angles for this rate, and mapping of stable and rock rubble areas based on inundation depths, topography, and soils. The WRC-Environmental and Swanson (1992) report also estimated slope failure cumulative frequencies, but the frequency assumptions are incorrect (RCE, 1993), so those results are not reviewed here. There has been no subsequent frequency analysis, and landslide magnitude-frequency relationships remain undetermined.

Long-term Hillslope Stability

The minimal impacts of slope failures estimated by the Sacramento District and the CDWR studies were also questioned on the basis that the time frame considered, 100 years, was too short (WRC-Environmental and Swanson, 1992). An understanding of long-term hillslope stability requires concepts of landform evolution, pedogenesis, and hydrologic, climatic, and vegetation change (Carson and Petley, 1970; Freeze, 1987; Brooks et al., 1993). Hillslopes evolve not only through a simple balance of instantaneous forces acting on isolated components of the system, but also through complex and nonlinear responses involving diverse factors such as delayed reactions to perturbations (Schumm, 1973, 1977; Graf, 1977; Cooke and Doornkamp, 1990).

Hillslope colluvium in the canyon is part of a sediment conveyance system where potential energy is maintained by channel erosion at the base. Processes that accelerate removal of material on lower slopes may oversteepen and destabilize upper slopes, accelerating transport from above. Propagation of instabilities may proceed through a series of delayed, complex, episodic, and indirect processes that can be hard to recognize, let alone anticipate. The importance of mass wasting in chaparral as an ongoing geomorphic process was long overlooked due to the infrequent but episodic nature of events. For example, the role of mass wasting to sediment budgets in chaparral was not appreciated until relatively recently (Bailey and Rice, 1969; Campbell, 1975; Rice, 1982). Prediction of slope failures is further complicated by site idiosyncrasies, including heterogeneity of vegetation and slope materials, fire histories, rates of debris recharge, and progressive failures. The link between fires and debris flows is well established (Rice, 1982; Wells, 1987). Unlike earthquakes, the duration of fire effects is substantial (several years), so the probability of the joint occurrence of inundation and fire effects in the canyon may be considerable.

The importance of hillslope stability in the canyon should not be underestimated, as extensive sliding may occur during single events in California chaparral environments. Both empirical and analytical evidence suggest that substantial slope instability and increased probability of slope failure could occur in the American River canyon in response to deep inundation and rapid drawdown, perhaps leading to increased mortality of vegetation (see next section), which in turn could be detrimental to long-term slope stability.

Plant Communities

Evaluation of the relative impacts and benefits of alternative flood damage reduction measures for the Sacramento area must consider the possible impacts in the Lower American River Parkway area, in the Sacramento Yolo Bypass, and on the native plant communities in the Auburn canyon area.

The difficulty in evaluating the dry dam alternative is that there is no comparable structure in a comparable environment from which reliable data can be collected to predict the environmental impacts of such a dam at the Auburn site. Reports completed in 1994 under contract to SAFCA and the Sacramento District concluded that existing inundation research and field observations at Auburn canyon and the Keswick Dam area of the Sacramento River do not adequately support defensible estimates of inundation effects on vegetation in the American River canyon (Chasse and Platenkamp, 1994; Hart et al., 1994). These reports concluded that the initial information developed for the 1991 ARWI and its environmental impact report and statement and habitat evaluation procedures was inadequate to support the quantified estimates of inundation losses used in the 1991 analysis. The reports provided constructive criticism of past research methods and made recommendations on how to design future research to arrive at quantifiable estimates that are more defensible.

Water project plans must attempt to quantify environmental impacts, not only in order to compare the relative merits of project alternatives, but also to design mitigation projects to replace lost ecological resources. Because little relevant information was available that could be applied to this case, most of the analysis has been necessarily speculative. The inherent contentiousness of this issue is evident in the history of reports and report reviews that have been undertaken thus far.

Estimating Plant Inundation Impacts

SAFCA concluded that the FWS study quantifying impacts on canyon vegetation, with its associated Habitat Evaluation Procedures analysis (FWS, 1991), produced an upwardly biased estimate of environmental impact. An alternative analysis commissioned by the California Reclamation Board and Department of Water Resources (Fugro-McClelland and Leiser, 1991), and released at the same time as the FWS analysis, arrived at lower estimates of quantifiable inundation impacts. A Sacramento-area expert who reviewed the Fugro-McClelland and Leiser report for the Planning and Conservation League (Jennings, 1991) and WRC-Environmental and Swanson (1992) raised significant technical issues about the research methods used in the report. In response, SAFCA also commissioned an evaluation of the Fugro-McClelland and Leiser report. In that evaluation, Keeley (1992) raised serious concerns about the extrapolation of data from existing research on deciduous floodplain species to the American River canyon environment, as was done in the Fugro-McClelland and Leiser report, and recommended that field experiments be conducted to determine a valid way to arrive at inundation-mortality relationships.

As a result of these challenges, the Sacramento District commissioned a study (Chasse and Platenkamp, 1994) to evaluate existing information available from research reports, including information from expert individuals in plant ecology and physiology, and from research specific to the American River canyon dry dam case. In addition, SAFCA contracted for some of the experimental field data proposed by Keeley. It hired consultants to conduct inundation studies involving the submergence of plants in Folsom Reservoir (Hart et al., 1994) and to collect field data from the Sacramento River below Keswick Dam that could be applied to the Auburn Canyon case (Meredith et al., 1994).

Experts now tend to agree that the research methods of the 1991 ARWI reports could not support quantifiable estimates of inundation impacts and that a research program dependent on more experimental field data on plant inundation, combined with a more thorough analysis of canyon geomorphological stability, should be conducted to provide scientifically defensible estimates. The Keeley (1992) review of the Fugro-McClelland and Leiser study on inundation impacts, the FWS report (substantiating report, Volume 1 (Appendix A-F) November 1991), the Planning and Conservation League report (Jennings, 1991), WRC-Environmental and Swanson (1992), Hart et al. (1994), and Chasse and Platenkamp (1994) agree that it is not reasonable to apply the existing research on plant inundation tolerances to Auburn canyon. These investigations found that the existing scientific literature on flood tolerance of plants refers mostly to bottomland, riparian, or riverside species. There is little or no information on the inundation tolerance of oak woodland, chaparral, and conifer forests such as those found in Auburn canyon. Ninety percent of the species in Auburn canyon are upland species. Of those, 74 percent are evergreen and, given the Mediterranean climate of California, have growing and dormant seasons that coincide with virtually none of the species on which information is available. In contrast to most deciduous species, these evergreens are physiologically active in the winter months, and many of them become dormant during the hot, dry summer period. The 8 percent of the canyon species located in the riparian zone, which are deciduous and typically go through a winter dormancy, can reasonably be compared to the flood tolerance data from the literature. The upland species cannot.

FWS made an extensive review of existing detention dams built by federal agencies in the West as part of its review of the 1991 ARWI. It could find no comparable bypass or dry dam situations in steep canyons with extensive acreage of upland-woodland or chaparral communities. FWS found that the dry dams that did exist in the West are substantially smaller and have significantly different ecological contexts. For instance, the plant communities in the Sacramento-Feather river bypass system were originally associated with a large inland sea and consisted of wetland and riparian species that could tolerate prolonged periods (up to several months) of flooding. Thus this system does not provide a good comparison because only a narrow corridor of the American River canyon contains riparian species. A report commissioned by USACE (Chasse and

Platenkamp, 1994) contributed to the FWS search for relevant plant mortality data using searches of various databases (e.g., Agricola, National Technical Information Services, and Waterways Experiment Station Dialogue). This additional search found that most of the literature reviewed was about ". . . oak woodland and montane forest species. No new citations were found that contained information on flood tolerances of upland species that would occur in the American River canyon" (Chasse and Platenkamp, 1994).

Experimental Research and Data Collection from Sites with Similar Plant Communities

Experimental field research involving the actual submersion of some Auburn Canyon plants in Folsom Reservoir and collection of observational data on plants affected by inundation on the Sacramento River were begun in 1994 in line with Keeley's (1992) recommendations to SAFCA. The plant submersion study (Hart et al., 1994) reviewed the 1991 ARWI and attributed the widely varying impact assessments to the lack of experimental field data.

The field experiment (Hart et al., 1994) entailed the lowering of some potted plants (a sampling of chaparral, oak woodland, and pine forest species) into Folsom Reservoir in February 1994, at depths up to 188 feet for up to 13 days. The results of this experiment may help to project potential inundation impacts and develop hypotheses on plant community shifts. Submergence killed certain chaparral species and damaged young tissue in all species. The statistical analysis seemed to indicate that all of the mortality and damage were due to the effects of duration, while none were due to depth. A statistically significant number of pine, live oak, coffeeberry, and manzanita plants survived submergence, but all showed some degree of damage or mortality. The study concluded that young plant individuals will probably be the most affected by inundation, and that evergreen shrubs and trees will experience some degree of damage or loss, with larger, mature individuals having greater rates of survival. The study also found that herbaceous plants will experience high mortality. There also may be a tendency in frequent inundation zones for a shift from evergreen to deciduous species. While a draft report of this field experiment was issued to the committee, the final results are still not available.

Both Hart and Keeley (personal communications, 1994) caution that there are significant differences in soil saturation and rate of drainage between pots and the natural watershed conditions present in the American River canyon. Potted plants will drain faster after inundation than soils in the canyon, which could remain saturated for longer periods. Soil saturation in the canyon is caused by a combination of inundation and rainfall, and could encourage the development of anaerobic conditions and subsequent growth of pathogenic fungi. Although experimental research such as this can be criticized for not simulating true field conditions, these studies at least bring insights close to realistic conditions.

A study on the Sacramento River floodplain below Keswick Dam was conducted in 1993 (Meredith et al., 1994), in which observations of plants affected by inundation in a March 1993 flood were made to understand more about the potential short-term effects on plant mortality, growth, and condition from a single event. These observations were combined with an analysis of long-term changes in tree and shrub cover in the inundation zone from 1966 to 1993, based on aerial photos and hydrologic records.

The aerial photos showed a 400 percent increase in canopy cover in one measured plot and a decrease in another plot of 35 percent. The researchers concluded that occasional flooding of varying durations and elevations does not eliminate natural recruitment of species in chaparral and oak woodland communities. The short-term mortality study collected data on 99 plants in the 1993 inundation zone. Mortalities included manzanita, live oak, and blue oak. Surviving species included coffeeberry, foothill pine, buck brush, western redbud, and mountain mahogany.

Reviews of the Keswick site study by Chasse and Platenkamp (1994) and Hart (personal communication, 1994) concluded that it has limited application to the American River because the site characteristics are considerably different and the Keswick study did not have appropriate controls for soils, geology, or hydrologic conditions to make comparisons clear. According to these reviews, the Keswick site is subject to the stress of considerable hydrologic scour. It has rocky soils, which are hard to saturate and drain quickly, and the site is more drought prone than the Auburn canyon. These conditions can either contribute to greater survival because good drainage or dryer conditions might be critical factors, or the conditions might represent stresses that produce higher mortality levels than would be the case in the upper American River canyon. Also, the vegetation is inundated as the Keswick site was flooded by fast-moving, turbulent water that is oxygen rich, as opposed to the deeper, more stagnant, oxygen poor water that would submerge the American River canyon plants. The lower oxygen potential in the root zones of the American River canyon plants is likely to be a cause of inundation induced mortality. Thus Chasse and Platenkamp (1994) and Hart (personal communication, 1994) concluded that the Keswick site study (Meredith et al., 1994) probably underestimates inundation mortality. Although the Keswick site study alone does not give directly useful data to support what would occur in the Auburn canyon, it will prove helpful where it supports the finding of other inundation studies (e.g., Hart et al., 1994) or where it supports future studies.

Projecting Indirect or Longer-term Impacts

The FWS critique of the 1991 ARWI raised the issue that inundating physiologically active plants will result in some level of stress that may be evident in several years, as opposed to creating immediate visible impacts. Keeley (1992) raised a concern over a potentially significant indirect impact of inundation of plant ecosystems. Studies show that the mycorrhizal fungi required for optimum plant growth in many species can be destroyed by short-term flooding. California shrub and oak species are known to be vulnerable to the loss of mycorrhizae. Other indirect mortality factors can include weakening of individuals and increasing susceptibility to disease or parasites. Beetle infestation as a secondary effect of inundation may, for example, be a contributor to existing unhealthy digger pines in the coffer-dam inundation zone. Observations of escaped exotics (nerium oleander) in areas below Keswick Dam and observations of invasive weedy species such as star thistle, annual grasses, and forbs in large sediment and slide deposits suggest potential impacts from invasive exotic plants after disturbance from inundation and slope slippage. None of the existing analyses have attempted to quantify or describe expected levels or ranges of risk from these indirect impacts.

Directions for Future Research

Attempts to predict mortality of vegetation have been contentious, but Chasse and Platenkamp (1994) have provided a valuable starting point for developing a research strategy that can hopefully garner confidence from a broader segment of the scientific community. The report summarized the findings of the different research documents prepared for the ongoing American River investigations, identified the areas of conflicts among the reports, and helped identify the points on which experts agree. The report stated unequivocally that the available information on plant inundation does not support precise estimates of inundation effects on vegetation in the American River canyon and called for a research strategy that openly acknowledges this uncertainty. The report then assimilated the data and recommendations from the existing reports and interviews with experts to identify the important data gaps, which should help structure plant inundation studies. The report acknowledged that it did not address the potential impacts from canyon landsliding and erosion, a factor that should be integrated with this effort to develop a comprehensive approach to future impact research.

The important data gaps summarized by Chasse and Platenkamp (1994) are as follows:

• Researchers have used estimates of inundation response for a dominant species (such as a canyon live oak) to characterize the response of a vegetation type (such as the evergreen hardwood woodland). However, the inundation responses of some of the most important species are not yet known, and it is likely that species within a vegetation type differ substantially in response.

• The reliability of extrapolations from inundation-mortality relationships of single species to those individuals in complex ecosystems is unknown.

• Most studies on the effect of inundation on tree and shrub mortality have

been observational rather than experimental, so little is known about the independent effects of duration, timing, and depth of inundation. Only by manipulating
depth, duration, and timing in controlled experiments will it be possible to make
such predictions.

• Vegetation maps for the project area are not yet adequate for use in
predicting vegetation loss under different inundation scenarios. The distribution
and acreages of dominant plant species should be mapped.

• The lack of data on vegetation regeneration in response to flooding makes
it difficult to project recovery of vegetation types or their conversion to other
types.

Because of the inherent difficulties involved in developing defensible, quantifiable estimates for inundation impacts, the committee recommends the formation of a team of recognized experts in plant physiology, ecology, and geomorphology to design a research program to follow up on the report by Chasse and
Platenkamp (1994). This program could combine strategies proposed by Keeley
(1992), Hart (1994), Chasse and Platenkamp (1994), and others involved in the
study of this issue. A practical combination of field observations, field experiments, vegetation mapping, modeling of landslide risks with different reservoir
operating scenarios, and landscape uniqueness evaluations could develop a credible environmental impact assessment. Developing ranges of estimates for inundation mortality or estimating minimum and maximum survival ranges (as Chasse
and Platenkamp (1994) have attempted) may provide the most defensible and
widely acceptable environmental impact analysis. At this point the geomorphological data suggest that soil slippage and slumpage may have at least as great an
impact on vegetation as inundation, so future investigations on impacts should
emphasize both these areas.

These data gaps and uncertainties dictate that certain precautions should be
included in the design and operating plan of any dry dam built at the Auburn site,
if one continues to be included among the flood control alternatives under consideration. The dam outlets should be designed to accommodate high sediment
loads that could occur with widespread slope failure. The gate design and operating policies must represent a compromise between minimizing frequency of
inundation and holding drawdown rates high enough to minimize mortality of
vegetation, but low enough to avoid substantial slope failure. These constraints
lead the committee to suggest that a dry dam be used only as a last resort, one that
would impound peak discharges from extreme events.

OTHER ISSUES OF CONCERN

Beyond consideration of the dry dam alternative, there are a number of other
environmental issues related to flood control planning in the American River
basin. These are generally less contentious than those surrounding the dry dam,

but nonetheless are important to highlight. They include impacts on riparian vegetation in the lower American River, the value of a geomorphic perspective, recreational conflicts, impacts on fish and wildlife resources, and the need for an ecosystem approach to environmental assessment.

Impacts on the Lower American River Plant Community

In addition to the environmental considerations assessed in the upper American River, the flood control planners also had to take a cautious approach to flood damage reduction alternatives that could affect the popular and heavily used American River Parkway on the lower American River, which flows through urban Sacramento. The alternative involving the increase in releases from a reoperated Folsom Reservoir through the American River Parkway levees was considered a contentious environmental issue because of projected impacts on parkway vegetation. Levee reconstruction and/or clearing for channel capacity or levee safety could have impacts on the riparian resources and quality of the river environment. The historically concerned and well-organized constituency associated with the parkway put the planners in the position of balancing public opinion of potential impacts to the upper American River against public opinion of potential impacts to the lower American River—a seemingly intractable position.

The *Detailed Report on Fish and Wildlife Resources* prepared for the 1991 ARWI by the FWS shows a net loss of 679 acres of riparian forest, marsh, and shrub vegetation along the lower American River for the without-dam 150-year protection alternative. This alternative would have changed the operations of Folsom Reservoir to release flows up to 180,000 cfs through the American River Parkway and levee system located in Sacramento. These estimates were arrived at assuming the need to rebuild American River levees, remove vegetation, riprap them, and remove vegetation to increase channel capacities in the floodplains. These impact estimates also assumed short- to long-term loss of riparian vegetation due to removal of the vegetation from the construction necessary to rebuild portions of the Sacramento Weir and Bypass and Yolo Bypass (Monty Knudsen, personal communication, FWS, August 1994).

New studies and institutional developments since the release of the 1991 ARWI reports change the impact assessments of this alternative on riparian and wetland species. The new institutional developments have served to relieve what was initially perceived as an untenable deadlock between upper American River public concerns and lower American River public concerns. As a result of innovative efforts by SAFCA to integrate the concerned public into its planning and design teams and of its openness to environmentally positive levee enhancement projects, the levee improvement projects can now be reclassified as environmentally beneficial projects. Levee improvement plans call for the integration of native riparian plantings into the projects, thereby providing net benefits for riparian plant communities compared to the existing situation (SAFCA, 1994b).

Other institutional changes include the previously mentioned program initiated by the California Resources Agency to view the Yolo Bypass as an opportunity for restoring and increasing riparian environments. In 1994 the agency announced new cooperative agreements to begin these restoration projects. The local and state agencies therefore are now viewing flood control improvement projects as opportunities to improve environmental values.

Geomorphological Influences on Flood Control

Water resources planning traditionally uses hydrologic data and hydraulic models as the focus of engineering studies. The realization that geomorphological influences, including stream dynamics, need to be routinely integrated into project designs and models has been advocated by fluvial geomorphologists for some time (Leopold, 1974), but only recently have hydraulic engineers and geomorphologists made progress in integrating consideration of natural river dynamics into project design and hydraulic models (Shields, 1982; Cook and Doornkamp, 1990; Neill et al., 1990; USACE, 1992).

The 1991 ARWI, for the most part, considered geomorphology in terms of its potential influence on upper American River environmental impacts associated with periodic flooding from a dry dam. But geomorphological issues should have received more consideration in the 1991 ARWI in analyzing the levee management options on the lower American River. Channel adjustment, which has the potential to increase the conveyance of floodwaters in the lower American, and the sediment transport and deposition in weirs and bypasses, are both important considerations not examined in the 1991 ARWI.

If geomorphological factors are not considered, options for increasing flood conveyance while limiting environmental impacts or increasing opportunities for environmental restoration can be overlooked. An increase in conveyance of floodwaters due to increased channel capacities (a result of channel degradation) may make it easier, for example, to allow more riparian restoration on the levees along the lower American River. An increase or decrease in storage capacities of bypasses also has implications for restoration opportunities. There may be systemwide benefits from reduction in flood damage if routine sediment removal at the Fremont and Sacramento weirs can improve the hydraulics of the Yolo Bypass and can lower water surface elevations upstream in the lower American. Sediment removal may represent a cost-effective and environmentally sensitive method of increasing lower river channel capacities.

An understanding of stream dynamics is critical to the design of levee improvements, particularly if there is a commitment to design the improvements with soil-bioengineering revegetation systems instead of traditional riprap. The anticipation of future channel adjustments becomes an integral part of multiple-objective levee improvement programs that must balance conveyance capacities and structural reliability of levees with riparian restoration opportunities. Finally, a better awareness of geomorphological processes provides a potential for incremental gains or losses of conveyance or storage capacities in the whole flood system.

Given the recent emphasis on considering more management options for the lower American, the local sponsor has commissioned consultants to evaluate the geomorphology in the lower American River as it relates to bank and levee stability. Given the geomorphological processes acting on the lower American River, the possibility of future channel degradation in the lower river deserves more consideration. The lower river may not have yet attained an equilibrium state from past historical influences. These two factors deserve attention in the formulation of current and ongoing alternatives.

Recreation Conflicts

Although the 1991 ARWI included nominal consideration and analysis of recreation resources and interests, it became clear in subsequent complaints that planners had not sufficiently involved this segment of the population in real negotiation during the formulation and evaluation of alternatives. Opposition to the dry dam proposal was to a significant extent organized by these interests. The SAFCA Lower American River Task Force is a step in the direction of resolving the impasse, but only a partial solution. Recreational issues will continue to be unresolved without substantial effort by all parties.

There are numerous recreational areas in the American River watershed that, by western standards, support heavy public use. For example, the American River is the most popular of all the white water rafting rivers in California (CSLC, 1994). The development and heavy use of these areas are due largely to the proximity of the basin to the dense population centers of Sacramento and the San Francisco Bay area. Recreation areas are located throughout the basin, from Discovery Park in Sacramento, up the lower American River along the American River Parkway, to the Folsom Recreation Area, and beyond to the Auburn State Recreation Area above the proposed Auburn dam site.

In the upper American River basin, the rivers act as natural corridors through the mountains and attracted human travel and activity long before contact with western civilization. The area is characterized by deep canyons with steep walls covered by chaparral, and narrow rugged valley bottoms and occasional rapids. White water rafting is a popular use of all three forks of the American River. The North and Middle Forks are particularly challenging, with many Class IV and V rapids, resulting in white water boating activities of state and national significance. The major rapids on the North and Middle Forks provide unique scenic features with minimal human intrusion. A dry dam in the upper canyon would significantly disrupt these activities and affect scenic and natural values.

The North Fork above the project area, from the Colfax Iowa-Hill bridge upstream to near Heath Springs, was designated a National Wild and Scenic River in 1978. In January 1993, the Bureau of Reclamation determined that the Middle Fork and the North Fork within the project area are eligible for Wild and Scenic designation, and a suitability study is under way.

The Auburn State Recreation Area lies mainly within the projected inundation zone of the originally proposed Bureau of Reclamation Auburn dam. The area is less than an hour from Sacramento and is visited by about half a million people each year. Because of its location and the diversity of opportunity, recreational use of this area will undoubtedly grow rapidly in the future.

In 1972 the lower American River was included in the State Wild and Scenic River System. In 1981 the exceptional anadromous salmonid fishery and other important recreational values of this reach of the river led to its designation as a unit of the National Wild and Scenic River System. The recreational units of the lower basin are linked together by an award-winning trail system. The Jedediah Smith Trail includes bicycle, pedestrian, and equestrian trails from Discovery Park to Folsom Reservoir.

Recreational facilities along the American River begin in Sacramento at Discovery Park, at the confluence of the American and Sacramento rivers. Above Discovery Park, the American River Parkway extends 23 miles upstream to the Folsom State Recreation Area at Nimbus Dam. The parkway is largely on the floodplain—bordered by high levees that isolate it from the surrounding urban development. The parkway functions not only as a recreational area, but increasingly as an urban transportation artery for pedestrians and bicycles. The parkway was used by an estimated 5.5 million people in 1988, and annual use is expected to grow to 7.5 million by year 2000 and 9.6 million by 2020. A 1983 survey found that more than half of these visits were associated with water-enhanced activities such as jogging, nature study, hiking, and picnicking, and that about a third of the visits were associated with water-related activities such as swimming, boating, and fishing. About 12 percent of the recreational use on the lower American River is by boating—primarily rafting, canoeing, and kayaking. These activities are highly seasonal in nature, with about 90 percent occurring between Memorial and Labor Days (USACE, Sacramento District, 1991).

Fishing continues to be the biggest recreational use of California rivers, and angling use of the lower American River is particularly important. About 55 percent of the total catch of chinook salmon in the freshwater of the entire Sacramento River basin for the year 1991 came from the American River. Catches of steelhead and American shad from the American River in the same year were also comparatively large, making up 48 and 44 percent, respectively, of the total Sacramento River basin harvest (CSLC, 1994).

Because of the size of Folsom Reservoir and its proximity to the Sacramento metropolitan area, Folsom State Recreation Area is one of the most heavily used areas in the state park system. The recreation area begins at the upper end of the parkway at Lake Natomas, an afterbay formed by Nimbus Dam. Recreational activities include fishing, power boating, sail boating, and windsurfing; there are conflicts between power boating and windsurfing. About 2.1 million people visit Folsom Reservoir each year, mostly Central Valley residents during the summer (USACE, Sacramento District, 1991). Reoperation of Folsom Reservoir will result in a lower pool during part of the year, adversely affecting recreational opportunities there. Mitigation is included in the reoperation plan.

Prior to urbanization and development, there was little public recreation development in Natomas, but bird hunting and watching on privately owned farmlands were common and continue today through the lease of hunting rights to hunting clubs. Much of the land along the Sacramento River in the Natomas area is privately owned, but the river channel is heavily utilized for recreational fishing and water sports including power boating, jet skiing, and kayaking. Development in the Natomas Basin would limit these recreational opportunities.

Impacts on Fish and Wildlife Resources

Given the importance placed on fish and wildlife by Sacramento area residents and visitors, potential impacts to these resources warrant careful review. Fish habitat in the project area of the North Fork has been degraded by a number of past actions, extending as far back as placer mining in the mid-1800s. Years of habitat degradation have combined with high summer water temperatures to limit the value of North Fork as a fishery resource in the reach that would be affected by a dam of any kind. Though the Middle Fork has also experienced some habitat degradation, the cool water outflow from Oxbow Dam supports a substantial population of large wild trout, both brown and rainbow. This population qualified as an "outstandingly remarkable" resource during the Wild and Scenic River Eligibility Assessment conducted by the Bureau of Reclamation in 1992.

The fish resources of Folsom Reservoir consist of both warmwater and coldwater species. The warmwater species, primarily bass, catfish, and sunfish, are adversely affected by fluctuations in surface elevation during the spawning season. These fluctuations, along with low nutrient levels in the reservoir, result in relatively low annual production for the warmwater fishes. The coldwater species, trout and salmon, are maintained by stocking, though limited natural reproduction occurs in tributary streams.

The once abundant chinook salmon resource of the Sacramento River basin has been reduced to a fraction of its original importance. Of the four distinct seasonal runs of this species, only the fall run now occurs in any numbers, and the winter-run fish is classified as endangered under the Federal Endangered Species Act (Fisher, 1994). Historic runs of salmon in the American River were estimated above 130,000 and included both spring- and fall-run fish. Both races of chinook were nearly decimated by hydraulic mining and dam construction in the late 1800s and early 1900s (Gerstung, 1971). The principal anadromous fish still surviving in the American River is the fall-run chinook salmon, now limited to the reach below Nimbus Dam. This population supports the extensive sport fishery mentioned above and also a significant sport and commercial harvest in the ocean. Over the period 1967 to 1991 (the baseline for the Central Valley Project Improvement Act), the river supported an average run of 32,000 naturally spawning fall chinook adults, about 22 percent of the total Sacramento River run of 143,000. Returns to Nimbus Hatchery below Nimbus Dam for the same period averaged 7,300 fish, 35 percent of the average Sacramento River total of 21,000 hatchery returns. The "naturally spawning" portion of the chinook run is actually heavily influenced by hatchery fish. Some of the fish spawning in the river are progeny of hatchery parents that fail to return to the hatchery, and some of the naturally produced fish interbreed with hatchery stock. The natural run in the American River has declined in recent years (the average run in the past 5 years was about 50 percent of the 25-year average). Steelhead in the American River are substantially less abundant and nearly entirely supported by hatchery production (1967 to 1991 average returns to Nimbus Hatchery were about 1,700 fish).

Fishery values in the Natomas Basin are much lower than those upriver. However, the Natomas area is highly significant for its wildlife values. Thousands of migratory waterfowl use the basin for feeding and resting. The Natomas basin reach of the Sacramento River supports one of the highest concentrations in California of nesting territories for the Swainson's hawk, a state-listed threatened species. And the southern portion of the American River basin in Sacramento and Sutter counties, including the Natomas basin, provides one of the most important habitats remaining in California for the threatened giant garter snake (EIP Associates, 1992).

Significant fish species in the zone of influence of the lower American River (principally in the San Francisco Bay and Delta, affected by flow releases from Folsom Reservoir) include the striped bass, which provides one of the most important sport fisheries of the state, and the endangered winter-run chinook salmon (reclassified from threatened status in January 1994), which passes the mouth of the American River on its way to spawning grounds in the upper Sacramento River basin. Other sensitive species affected by American River flows include the federally listed as threatened delta smelt, primarily resident in the bay and delta, and the Sacramento splittail (proposed for threatened status by the FWS in January 1994), which occurs both in the delta and in the lower reaches of the American River.

Two other federally listed species occur in the project area, the bald eagle (federally listed as threatened) and the valley elderberry longhorn beetle (federally listed as threatened). The eagle occurs in significant numbers only on Folsom Reservoir. The beetle occurs in association with elderberry shrubs primarily in riparian areas of the upper canyon and the lower American River.

Owing to the extent of historical habitat degradation in the upper basin, impacts of a detention dam in the canyon are much more likely to be significant to recreational and rafting interests than to fisheries or fish habitat, particularly in the North Fork. However, the important wild trout fishery in the Middle Fork could be substantially affected if canyon wall sloughing following inundation is extensive.

Under current operations the major limitation to success of the fishery resource of the lower American River is the flow and temperature regime below Nimbus Dam. The period of major concern is during the spawning migration of the fall-run chinook salmon. Owing to low flows during that time and to inadequate control of temperature of the releases from Folsom Dam, temperatures in the river often exceed those suitable for survival of incubating salmon embryos. Also of concern are temperatures and flows for rearing of juvenile salmon and steelhead during spring or summer. The salmon are less at risk because the juveniles leave the river by the early summer of their first year, before temperatures reach maximum levels. Steelhead, however, must rear one or two full years in the river before moving to the ocean. As a consequence of high summer temperatures and limited flows, natural rearing of steelhead has been virtually eliminated; more than 95 percent of returning fish are the result of hatchery rearing (Snider and Gerstung, 1986). A number of State Water Resources Control Board rulings regulate releases from Folsom Reservoir but they are inadequate to protect fish habitat. High temperatures and substantial and rapid fluctuation in flows are a major limitation to significant natural production of salmonids in the lower river (Snider and Gerstung, 1986; Williams, 1995).

Some relief is potentially available as a result of a recent court decision, resolution of which is still evolving. The case involved the *Environmental Defense Fund et al. v. East Bay Municipal Utilities District et al.* At issue was where EBMUD would be allowed to divert an annual 150,000 acre feet, for which it had contracted with the Bureau of Reclamation. The utility district wished to divert the water through the Folsom-South Canal, above Nimbus Dam. Environmental groups and others held that the diversion should occur lower down in the river system to protect the public trust resources of the river. In a decision handed down in January 1990, Judge Hodge of the Alameda County Superior Court allowed diversion through the Folsom-South Canal, provided that sufficient flow was available in the lower American River to support the anadromous fishery and other trust resources. The judge approved minimum flows for each season and mandated an ongoing research program. He also appointed a Special Master to oversee the research, which was to be directed toward reducing the overwhelming uncertainty that surfaced throughout the trial and also toward more accurately defining the required minimum flows (Williams, 1995). The decision was based in part on the Public Trust Doctrine (Sax, 1993) and has the potential to influence water management in the state for some time. Owing to provisions in the state constitution, members of the public in California have a special right to use navigable waters for all purposes. The Public Trust Doctrine gives the state particular responsibilities for protecting all beneficial uses of such waters (CSLC, 1994), and it was this authority, in part, invoked in the Hodge decision.

Given that reservoir operation already has a substantial detrimental impact on the fish populations, there seem to be no significant additional impacts on fisheries of the reservoir or the lower river from any of the alternatives in the original 1991 ARWI proposal. It also appears that no major additional effects would be associated with interim reoperation of Folsom Reservoir. In fact, the reoperation EIR/EA (SAFCA, 1994a) made several significant concessions to the anadromous fishery and to protection of endangered species. It ensures that if reoperation would require flow levels lower than the "Hodge flows," then Hodge flows would be met, provided that water were available. This obligation would be met by converting, to the extent possible, all potential environmental impacts to reductions in CVP water delivery. In a contract with the Bureau of Reclamation, signed in March 1995, SAFCA agreed to compensate the federal government for this water debt by acquiring sufficient water or water rights from other sources. SAFCA also agreed to finance modifications to the temperature control louvers in Folsom Dam to ameliorate high temperatures in the river, and to fund an evaluation of the impacts of reservoir reoperation on habitat of the Sacramento splittail in the lower river (SAFCA, 1994a).

Direct impacts of flood reduction measures in the Natomas Basin appear to be insignificant, but the indirect impacts of the additional development that would be allowed by flood protection could be important to habitat of the threatened giant garter snake and Swainson's hawk. The giant garter snake, listed as threatened by the state and federally listed as threatened in October 1993, has recently been given a high profile by the National Biological Service. Development of a giant garter snake Habitat Conservation Plan has been named one of 12 new national priority ecosystem initiatives of the agency. Though contractors for SAFCA had already produced a draft Habitat Conservation Plan for both the Swainson's hawk and the giant garter snake (EIP Associates, 1992), the plan to be developed under this newer proposal will be critical to land development plans in the Natomas Basin.

Another major source of uncertainty in the realm of aquatic resource issues is the question of how the additional water required to reduce salinity and improve habitat for endangered fish species in the bay and delta will be allocated. The complicated and interwoven set of circumstances surrounding this issue is discussed in Chapter 6.

Projecting Ecosystem Responses in Impact Assessments

Since the preparation of the 1991 ARWI reports, federal resource management agencies have adopted new strategies to consider and evaluate potential impacts within the framework of whole ecological systems. This shift is an effort to correct for past practices. Too often in the past, the focus of environmental assessment has been on dominant plant or animal species at the expense of understanding the important role that interactions among species and their environment may have on the species and community survival and the role of corridors and linkages of natural environments. The 1991 ARWI report was, for the most part, no exception to this narrow focus, although the FWS did raise the possibility of significant shifts in ecological systems due to disturbances that a dry dam could trigger. It noted the absence of information on impacts to the plant *communities* (emphasis added).

Ecosystem responses to a dry dam could include wildlife community shifts associated with plant community shifts because of habitat changes. Positive ecosystem changes could occur to the Lower American Parkway, in which levee rebuilding and associated revegetation projects could help reintroduce greater riparian species diversity.

Because of the significance of potential impacts of inundation on the plant communities in the American River canyon, an ecosystem framework for description of these probable impacts is particularly important. An adequate environmental assessment should attempt to provide descriptions on how ecosystem dynamics, function, and structure could react to changes made to the system. In the 1991 ARWI report the focus on potential inundation tolerances of individual species loses sight of this critical larger picture. The potential for ecosystem shifts in the American River canyon could be related to the direct impacts of periodic inundation on plants or to the indirect impacts previously discussed, such as changes to the composition of soils, soil microbiota, or community tree or shrub densities.

The significance of such potential ecosystem losses and shifts needs to be discussed in a regional context. An example of a regional ecosystem approach to characterizing environmental impacts would be a discussion of the regional or statewide value of riparian and oak woodlands. The 1991 ARWI noted that of the state's original riparian habitat, less than 5 percent remains today. Moreover, less than 2 to 3 percent of the woody riparian habitat remains along the Sacramento River. What is the value of the riparian environment in the upper American River in this context? What is the value of the upland woodlands that could be lost through a combination of inundation and hillslope failure?

An effort to frame impacts in a regional context could, for example, recognize that oak woodlands are an ecosystem of increasing concern to plant community ecologists. Studies of foothill oak populations indicate that they do not have the age distribution of healthy vigorous populations. Communities of valley, blue, and Engelmann oaks show a narrow cluster of middle-aged trees, with few young or old ones. The fear is that middle-aged oaks could reach the natural limit of their life span and disappear, to be replaced by other less productive and less diverse communities. Lack of reproductive success is attributed to a combination of factors including deforestation, overgrazing by gophers, deer, and cattle, introduction of exotic grasses, and alteration of fire cycles. It is estimated that the state has lost over a million acres of oak woodland since the 1940s (Barbour et al., 1993). The potential area of impact in the upper canyon support two of the state's three "oaks of special concern," the valley oak, *Quercus lobata*, and the blue oak, *Quercus douglasii*.

Although sensitive species of oaks could be of concern, inventories of the Cosumnes River watershed suggest that rare landscape forms could be of even greater regional significance in the upper American River watershed. Inventories of rare landforms and vegetative patterns indicate that montane oak woodland and shrub communities are more widespread, for example, than the remnants of Central Valley riparian systems and associated oak woodlands. Nonetheless, a central issue to explore is, how many miles of canyons with free-flowing rivers, scour zones, steep canyon walls, and a diversity of ecosystems do we have (Hart, personal communication, 1995)? Future impact evaluations should draw on the methods for inventorying and evaluating landscape uniqueness (Leopold, 1969; Riley, 1974).

Scenarios should be developed for the potential ecosystem shifts of each vegetative community that could be affected. Information so far suggests the loss of old and young individuals and shifts to deciduous species. Chaparral communities disturbed by periodic inundation or landslides could shift to grasslands or even "communities" of invasive exotics. Inundation would likely result in the replacement of any native perennial grasses, ranked very rare by the state, by more weedy, nonnative annual grasses. Disturbances in lower canyon elevations could increase the extent of riparian zones, which typically occupy disturbed environments (Knudsen, 1991; Keeley, 1992; Hart et al., 1994; Meredith et al., 1994). While the forecasting of ecosystem shifts still remains in the realm of speculation, the potential for both positive and negative ecological and aesthetic changes should become part of the evaluation of impacts in future assessments.

CONCLUSION

The 1991 ARWI raised a number of contentious environmental issues, including debate over whether the descriptions of the environmental impacts of the various proposed alternatives was adequate. Based on its review, the committee determined that there were several areas of the 1991 ARWI where the lack of scientifically-based descriptions of environmental impacts prevented the report from serving as an adequate planning document. The most significant deficiencies were in the assessment of impacts that might be caused by periodic inundations from a dry dam on the plant communities in the upper American River canyon, the impacts of inundation on canyon soils and geologic stability, and an ecosystem and regionally-based assessment of impacts. Whether subsequent activities are filling some or perhaps all of these gaps is unclear at this time, but these questions should be resolved with the expected publication of the Sacramento District's Draft Supplemental Information Report, expected in the summer of 1995.

Public officials ultimately face a difficult decision: given the significant flood hazard to Sacramento, landslide hazards in the American River canyon may be deemed a necessary cost of flood protection. If this option is pursued, great sensitivity to environmental values should be incorporated. If a dry dam continues to be included among the alternatives under consideration, the committee suggests that the following research needs and issues be given consideration:

• The Sacramento District should form a team of experts in plant physiology, plant ecology, and geomorphology to design a research plan that combines field experiments, observations, vegetation mapping, landscape uniqueness data, and modeling of landslide risks to develop a canyon inundation impact assessment that can secure acceptance and credibility from the scientific community.

• The two large old slides should be thoroughly mapped, analyzed, and monitored to assess the potential hazards of catastrophic failure.

• If dry dam outlets and storage are to be used, they should be designed to accommodate high sediment loads in anticipation of a worst-case scenario of numerous hillslope failures.

• Rates of drawdown should be minimized but should be sufficient to prevent substantial vegetation mortality until more is known about long-term vegetation responses to inundation and slope responses to subtle vegetation changes.

• Gate design and operating policies should consider the depth and frequency of inundation while keeping drawdown rates low. These conflicting constraints prevent a dry dam from being used as a first line of defense and instead restrict it to use as a last resort to contain peak discharges from extreme rare events. This philosophy could be made explicit in the Folsom Dam operating policy.

4

Risk Methodology

The U.S. Army Corps of Engineers (USACE) has adopted new risk and uncertainty analysis procedures for project evaluation that explicitly include uncertainties in the hydrology, hydraulics, and economics of a planning study (USACE, EC 1105-2-205, 1994) (hereafter referred to as EC 1105-2-205). This procedure represents an extension of the traditional paradigm for flood control project planning and community flood protection evaluation. USACE observed that the new risk and uncertainty methodology is similar to present practice but differs in that uncertainty is explicitly quantified and integrated into the analysis (USACE, EC 1105-2-205, 1994).

The 1994 Alternatives Report (USACE, Sacramento District, 1994a) indicated that USACE's analysis now considers "varying degrees of uncertainty in the causes of flooding, such as inflow to Folsom Reservoir, regulated outflow-frequency relationships for Folsom Dam, river stages, and levee stability." The methodology computes the risk of flooding due to combinations of hydrologic events, hydrologic parameter uncertainty, uncertainty in stage-discharge relations, and levee performance.

This change in methodology is important to the American River Watershed Investigation (ARWI) because the ongoing evaluation of flood control alternatives for the basin by the Sacramento District is one of the first applications of the approach, and almost certainly the most complex application yet attempted by USACE. The risk and uncertainty methodology specifically addresses many assumptions in the 1991 ARWI that were subject to controversy, and which the committee was charged to review. Whether the controversy will be resolved remains to be seen.

In particular, assumptions about levee freeboard for American River basin levees are replaced by a distribution on the stage at which levees fail. Likewise, hydrologic uncertainty that was described by an expected probability adjustment, and assumptions about delays between the beginning of the flood and increased releases, are now described by explicit probability distributions. Issues that were in contention have not disappeared; what some viewed as conservative values have been replaced by probability distributions, which may also be contested.

For decades, civil engineers have realized that it is not practical to protect communities in the floodplain from all conceivable floods (Foster, 1924; Riggs, 1966). Such protection measures would be prohibitively expensive, even if they were practicable. Communities and individuals who choose to locate in flood-prone areas will generally be exposed to some risk of flooding. However, it is often economically advantageous to provide protection from flood events that have a 1 in 50, 1 in 100, or a 1 in 500 chance of occurring in any year, depending on the value of the property at risk, the chance of loss of life, and the costs of flood risk reduction opportunities. Derivation of probability distributions to describe the possible magnitude of flood flows has been a practice in civil engineering since the early part of the century. They provide a description of hydrologic risk. When a particular flood flow with a 1 in T chance of being exceeded in any year serves as a design flood for a project, USACE has said that the project provides a T-year "level of protection."

The new USACE risk and uncertainty methodology explicitly introduces into the planning process consideration of hydrologic, hydraulic, and economic uncertainty. Before, the USACE planning procedure selected a level of protection corresponding to perhaps the 1 in 250 chance event (often called the 250-year flood), and then determined the corresponding design flood flow. Use of an expected probability correction did incorporate hydrologic uncertainty into flood risk estimates (Beard, 1960, 1978; Stedinger, 1983a). Alternative hydraulic flood control structures including levees, flood storage capacity in dams, and channel improvements, in addition to flood-proofing efforts, were selected to control a flood of that magnitude.

In the evaluation of flood control projects, there are a number of uncertainties that make it difficult to determine whether a specified flood can be passed safely. For example, flood control dams might have surcharge capacity that was not included in the flood routing calculations. Levees are a more common concern. Levee failure depends on factors such as the structural integrity of the levee embankment, possible scour and undercutting, variation in the state of levee repair, and other factors in addition to high water levels. Hydraulic predictions of the flood stage associated with different flow rates may also be in error. Levee failure stage predictions and stage-discharge relationships are affected by surveying inaccuracies in the measurements of channel geometry and riverbed elevations, errors in estimation of flow resistance, simplifications in hydraulic routing calculations, waves and wave effects, and possible settling of levees that affect crest elevation.

Risk-based analysis of hydrologic and hydraulic engineering problems has been and is an active area of research (Davis et al., 1972; Tung and Mays, 1981; Haimes and Stakhiv, 1986, 1989, 1990; Duckstein et al., 1987; USACE, 1992a,b; Haimes et al., 1993; Taylor et al., 1993). In most risk analysis applications, the risks of concern arise from the distributions of annual flood peaks, rainfall depths, and other hydrologic variables (Mays and Tung, 1992). In a few cases, project performance is described probabilistically (Duckstein and Bernier, 1986; Chow et al., 1988, section 13.4; Mays and Tung, 1992). Uncertainty in structure performance was important in several studies addressing dam rehabilitation and dam safety (McCann et al., 1985; Goicoechea et al., 1987; Von Thun, 1987; Stedinger et al., 1989; Bowles, 1990; see also NRC, 1985).

There are relatively few applications where risk analyses have considered the natural variability in hydrologic and hydraulic variables as well as the uncertainty in the parameters of fitted flood-flow frequency curves and calculated stage-discharge relationships, and in economic quantities; these analyses might best be described as risk and uncertainty analyses to make the distinction clear. The Bayesian[1] framework that is appropriate for hydrologic uncertainty has been employed in proposals to include hydrologic parameter uncertainty in planning studies (Benjamin and Cornell, 1970; Duckstein et al., 1975; Vicens et al., 1975; Wood, 1978; Stedinger, 1983a). The USACE use of expected probability adjustments is one way to include parameter uncertainty in flood control project evaluation.

RISK AND UNCERTAINTY: A PRIMER

Uncertainties, Safety Factors, and the Meaning of Level of Protection

USACE traditionally has included safety factors in its design of facilities and the specification of operating policies to address important hydraulic uncertainties in flood control planning calculations. Surcharge storage in reservoirs might be one safety factor. For levees, engineers have required that the design flood

---

[1]The statistical literature includes several methods for dealing with parameter estimation, statistical inference, and decisionmaking. Bayesian statistical methods treat unknown statistical parameters (the population mean, population variance, or a probability or quantile) as random variables whose probability distributions reflect the degree to which the value of a parameter can be resolved from available sample information as well as prior beliefs and other sources of information. With the traditional statistical procedures employing standard confidence interval estimators and classical hypothesis tests, such parameters are treated as if they have fixed (but unknown) values, and probability distributions describing the sample-to-sample variability of sample statistics and parameter estimators are the focus of the analyses. The topic is addressed in more detail in the text.

pass through the levee system with some specified freeboard. Such a safety factor enables the engineer and the planning agency to be confident that in an actual flood event approximating the design storm, there will be sufficient channel capacity to pass that flood flow without the levees failing from overtopping or excessive stages. In planning studies, encroachment within levee freeboard might be treated as sufficient to cause levee failure, even though in an actual flood failure might not necessarily occur at that stage. From an economic perspective, one can ask how much freeboard is justified economically to increase project reliability (Davis, 1991).

The practice of including freeboard in design suggests that at the design flood associated with a target probability, called the "level of protection," there will often be some residual safety factor before actual flooding would occur. If there is, then the true chance of levee failure resulting in major flooding is less than the specified target probability. The question arises as to what was meant by the traditional "level of protection." Should it have been viewed as (1) an estimate of the chance of flooding due to levee overtopping or breaching, or was it simply (2) the exceedance probability of the design flood that a reservoir and levee system was designed to pass with some safety margin?

Generally, evacuation plans would begin before a levee breached or was overtopped. Thus the "level of protection" could be viewed as the probability that the design event would be exceeded and thus that emergency measures would be required, even though widespread flooding might not occur.

What seems clear is that there is confusion on this issue. Although calculated levels of protection might appear to address (2) above, their use to estimate expected damages suggests that they are often used as an answer to (1). This has led to the conventional wisdom that USACE projects provide more protection than acknowledged because safety factors built into levee design and reservoir operating policies appear to add an additional increment of safety. If this conventional wisdom is true, then by lowering the apparent benefit-cost ratios this practice may have worked against some proposals to provide needed flood protection. For example, if levees can almost always pass flood flows that encroach within the specified design freeboard, they actually provide protection from larger floods than has been assumed in many analyses.

However, the inclusion of safety factors in reservoir-levee system design to compensate for hydraulic uncertainty may not be sufficient to actually decrease the risk of levee system failure or levee overtopping. If levee settlement in one location ensures that a levee system failure will occur before the design flood event is reached, excess channel capacity or extra freeboard at other locations will not improve system reliability. In a levee system, failure occurs at the weakest point. However, if in a flood event a reservoir operator can vary releases in response to actual developments within the channel-levee system, it is possible that variation in reservoir operations taking advantage of excess surcharge storage could avoid levee system failures due to other deficiencies within some range of hydrologic loading.

Planners and engineers also realize that the condition of levees and some equipment degrades with time. Safety factors are a reasonable way for designers of flood control works to ensure that over time a system can continue to pass the design flood without levee overtopping or breaching. However, it may not be immediately clear how safety factors included in different components of a reservoir-channel-levee system interact to affect overall system reliability.

Definitions for Risk and Uncertainty

USACE will be wrestling for some time with the implementation of its new risk and uncertainty methodology. Of concern will be both a consistent scientific methodology, and a vocabulary and style for the presentation of the results to technical audiences and the public. The choice of words is very important because they help us distinguish one concept or idea from another. In this regard, the terms "risk" and "uncertainty" can cause problems because different authors have ascribed to them significantly different ideas. Risk has been used to convey each of the following meanings (USACE, 1992a, pp. 10-11):

1. The idea of hazard, when something is described as being "at risk."
2. The expected losses or risk related to a venture.
3. The probability of some outcome, such as the risk that a levee will be overtopped.

All three definitions attribute to risk a probabilistic character related to the possibility of an adverse and unwanted event in a particular system. Risk may be due solely to physical phenomenon or to the interaction between man-made systems and natural events.

The term uncertainty has been given a broad and sometimes conflicting range of meanings. There is a literature wherein the term uncertainty is used to describe events for which objective probabilities are not available (USACE, 1992a). On the other hand, it could simply to be used to describe situations that are not certain; USACE (1992a) stated that "uncertainty means simply the lack of certainty. It is the reality of inadequate information. When information is imprecise or absent, that is uncertainty."

The USACE's guidelines provide the following operational definitions of risk and uncertainty (USACE, 1992a, p. 12):

> Risk: The potential for realization of unwanted, adverse consequences; estimation of risk is usually based on the expected result of the conditional probability of the occurrence of event multiplied by the consequence of the event, given that it has occurred.

Uncertainty: Uncertain situations are those in which the probability of potential outcomes and their results cannot be described by objectively known probability distributions, or the outcomes themselves, or the results of those outcomes are indeterminate.

Those guidelines indicate that actual uncertain planning situations are located on a continuum between situations of known risk (where the probability distributions of interest are well specified) and situations characterized by uncertainty (where those distributions are hardly specified at all; USACE, 1992a).

A Distinction Between Risk and Uncertainty

Although the cited distinctions between risk and uncertainty are some times useful, they are not the distinctions that are needed for our discussion of the USACE methodology for risk and uncertainty analysis. Of particular concern here with regard to the USACE risk and uncertainty methodology are:

- models of natural and operational variability and randomness, including probability distributions describing flood flows, event-to-event variability in stage-discharge relationships and reservoir operations, and variability in flood damages due to factors not captured by flood stage, and
- uncertainty representing limited understanding of system processes and the lack of accuracy with which the parameters in models describing natural processes can be specified, including the parameters of a probability distribution, the cross-sections used to derive a stage-discharge curve, and the value and the count of the number of dwellings in a protected portion of the floodplain.

In some cases one may be uncertain as to which of several competing models to employ, such as alternative probability distributions. Uncertainty refers to our lack of understanding of characteristics of nature that we conceptualize as being fixed at any given time. Ideally, the values of various model parameters could be determined. However, due to data limitations there are generally residual errors in our understanding of those characteristics of nature that cannot be eliminated with reasonable levels of effort.

The first situation is referred to here as natural variability or randomness in the indicated process. The second situation is referred to as "specification error," or simply uncertainty. This use of uncertainty to describe lack of knowledge is not strictly consistent with the operational definition for the term suggested in USACE (1992a), although it may be consistent with the way the term is used. This use is consistent with the definitions adopted by other groups (ISO TAG 4, 1993; Taylor and Kuyatt, 1993; NRC, 1994).

Sources of Uncertainty

Recently, in a report on risk assessment of hazardous air pollutants, the National Research Council (NRC, 1994) recommended making a clear distinction between parameter uncertainty, which is associated with the parameters of a particular model, and uncertainty as to the appropriate model, or model uncertainty. The report noted that parameter uncertainty often is described by continuous parameter ranges (NRC, 1994) that result in corresponding uncertainty intervals associated with predictions; however the choice among competing health risk models generally corresponds to distinct and mutually exclusive choices. The authors observed that "indiscriminately" combining the two types of uncertainty in health risk assessment could result in the calculation of average health risks and uncertainty ranges that are inconsistent with any of the alternative models. The report recommended that parameter uncertainty be evaluated separately for each competing model. Hydrologists face similar issues when choosing between alternative flood flow probability distributions or between methods for calculation of stage-discharge relations.

Hydrologists often consider what can be classified as a third type of uncertainty, which arises due to model imprecision, or model prediction error. Thus, even with the best parameters, operational hydrologic models may fail to precisely predict flood stages at some locations in a system; such model prediction errors are another source of uncertainty in the analysis of flood projects. The error here is not due to natural variability, which might be best described explicitly, or to a failure to have the best set of model parameters, which is described by model parameter uncertainty, but is instead due to lack of model accuracy and thus is a source of uncertainty associated with model predictions. Such prediction errors can be thought of as a type of model uncertainty, because if one had a more accurate model, such errors might be eliminated. However, better models in most cases would have greater data requirements, requiring a finer spatial description of channel cross-sections and roughness coefficients with fewer lumped representations of watershed and channel characteristics. In fact, most operational models deliberately employ simplifications and lumped representations of natural processes to restrict the parameter space to a manageable dimension so that available data are sufficient for model calibration. Thus uncertainty due to model prediction error often reflects both data/parameter limitations and model uncertainty. In this discussion, model prediction error is included with other parameter and model uncertainties.

A FRAMEWORK FOR RISK AND UNCERTAINTY ANALYSES

A framework is needed to understand the structure of risk and uncertainty analysis efforts for flood protection project evaluation and to understand the relative roles of the natural variability of flood volumes, reservoir operations,

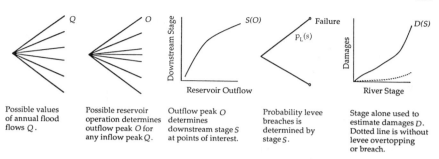

| Possible values of annual flood flows Q. | Possible reservoir operation determines outflow peak O for any inflow peak Q. | Outflow peak O determines downstream stage S at points of interest. | Probability levee breaches is determined by stage S. | Stage alone used to estimate damages D. Dotted line is without levee overtopping or breach. |

FIGURE 4.1 Deterministic and stochastic processes contributing to flood risk. Performance of a flood control system involving both reservoirs and downstream levees can depend on deterministic and stochastic components. Possible values of the inflow peak for any year are described by a tree with branches, as are reservoir operations during that event, because both are described as random processes. The transformation of the outflow peak O to downstream stage S is described by a deterministic relationship, though there is uncertainty associated with parameters of that relationship. Likewise, damages are described as a deterministic function of river stage for the levee breach/overtopping case, and the case without levee failure or over topping. Levee failure is probabilistic and occurs with a probability $p_L(S)$, which depends on the stage S.

hydraulic system performance, stage-discharge errors, and uncertainty in hydrologic, hydraulic, and economic parameters. Figure 4.1 provides a conceptual model for describing hydrologic risk, variation in reservoir operations, use of a river stage–reservoir outflow relationship, levee reliability, and finally estimates of the economic damages that would result should a levee fail. Several of these relationships are stochastic, while others are described by deterministic relationships.

The committee developed the event tree in Figure 4.1 to describe how the volume distribution of the largest flood volume in a year is transformed first into a river stage distribution and eventually into a damage distribution. This event tree can be used to evaluate the probability that flood protection works are overwhelmed and flooding occurs at some damage site, called the annual failure probability (AFP). Likewise, it can serve as the basis for calculating the expected annual damages (EAD), which would be the foundation of the economic evaluation of project performance.

In Figure 4.1, a process is modeled either as being deterministic or as having some random component reflecting natural process variability. To understand the impact of specification errors or uncertainty in parameters of the selected discharge-frequency model or in economic parameters, it is useful to note that for each step in Figure 4.1 there is a set of parameters that define the relationship or model employed at that step. For example, in the first step the flood flow frequency relation requires specification of the parameters of that distribution;

these are often taken to be the mean, variance, and skewness coefficient of the logarithms of the flows. Likewise, the variability in reservoir operation at the second step will be described by some selected probability distribution, which will also have parameters. In this presentation, uncertainty analysis focuses on the parameters of the selected models; those models are assumed to accurately reflect the probability distributions of processes that are variable (such as the largest inflow in a year and the actual timing of reservoir releases in a future flood event) and of deterministic processes such as the stage associated with different channel flow, if only the correct or best values of the models' parameters could be determined.

When the problem is structured as it is in Figure 4.1, one can identify the parameters of each of the models that determine the numbers that enter into calculation of risk and expected damages. One might then ask, how well or how precisely are those model parameters defined? Or, how uncertain are values of the project performance criteria AFP and EAD owing to the uncertainty or specification errors in various parameters?

There are several sources of variability in the economic damages that will be experienced in any year. Extreme variability results from the magnitude of the floods that may occur. Less important but still significant variability is introduced by flood hydrograph timing and shape, variations in reservoir operations, possible levee failure stage, and differences in the actual damages that would occur to a structure depending on the duration of flooding, wave attack, and differences in warning times; the effects of these factors are not captured by the specification of stage alone. Planners understand that this variability exists and so base their plans on AFP and EAD, which reflect the decision to average over the probability distributions describing annual maximum flood volumes and other variable processes.

In structuring the problem, as has been done in Figure 4.1, engineers can also clarify how the various processes are thought to work. For example, the stage-discharge relation can be conceived of as being time-invariant or deterministic, so that a specific stage always corresponds to the same discharge. Then the relevant uncertainty would pertain to the precise functional relation between discharge and stage. Alternatively, there are certain stream reaches where the stage-discharge relation varies significantly over time because of channel changes, sediment movement, or the stages of tributaries or other streams with which the river of interest merges. Such stage-discharge relations hence might best be described by some random process. While in this second case the stage-discharge relation might best be described as a source of variation, there would still be uncertainty as to the best values of the parameters that describe that process.

Economic damages depend on several factors, and some are deterministic while others are random. In particular, actual flood damages vary depending on flood duration, the presence of ice and sediment, wave action, and warning time. Flood damage uncertainties related to the number, types, and value of structures in flood-prone areas would not change much from year to year, unless a major flood occurs. The source and nature of variability and uncertainty in levee performance present similar issues. USACE needs a clear framework for its risk and uncertainty calculations to be able to articulate and explain its treatment of such issues. Even so, it will not always be clear what should be described as variability and what to represent as uncertainty.

Including Uncertainty in the Analysis

Planners should know by how much the estimates of AFP and EAD might be in error. For example, a flood-frequency curve is based on a limited flood record. By how much might the parameters of the discharge frequency function be in error, and how big a change in AFP and EAD would result? Likewise, in determining the stage-discharge relationship, a limited amount of effort goes into the surveying and the description of river cross-sections, geometry, and roughness coefficients: the hydraulic model has a limited amount of detail. What errors might this introduce into the evaluation of AFP and EAD? Similarly, limited effort is devoted to determining the value of property at risk in flood-prone areas. Additional effort could refine the data base describing the property at risk. Given a statistical description of the likely specification errors in economic and structural survey data, a planner could quantify the magnitude of the corresponding errors in AFP and EAD.

These questions can be addressed by sensitivity analysis procedures. The document *Guidelines for Risk and Uncertainty Analysis in Water Resources Planning* (USACE, 1992a), developed by the USACE Institute for Water Resources, defined sensitivity analysis as

> the technique of varying assumptions to examine the effects of alternative assumptions on plan formulation, evaluation and selection. This can include variation of model parameters as variation of benefit, cost and safety parameters. One of the important uses of sensitivity analysis is to investigate how different values of certain critical assumptions and parameters could result in changing the choice of the selected project and report recommendations. Sensitivity analysis is the systematic evaluation of the impacts on project formulation and justification resulting from changes in key assumptions underlying the analysis.
>
> Sensitivity analysis can be used to bracket forecasts, parameters, benefit and cost estimates, and other factors for which a range of values can be expected to occur.

Generally, each model or process parameter is varied, one at a time, and the result observed (USACE, 1992a). However, there are often so many parameters in the models employed to evaluate flood protection projects that it would be difficult to integrate such one-at-a-time evaluations, or to decide how they should be incorporated into decisions (Moser, 1994).

Describing Uncertainty

Useful descriptions of uncertainty can be developed by describing the specification errors or uncertainty in various economic parameters by probability distributions. This must be done with care so that the resultant distributions truly reflect the probabilities planners should ascribe to the various parameters given the sample information at their disposal, general information they have about the processes of interest, and what is reasonable for the location in question. Then, using those probability distributions over the uncertain parameters, a statistical description of the uncertainty in AFP, EAD, and other performance criteria can be computed.

For the purpose of developing a more mathematically precise notation for describing uncertainty, denote a possible set of model parameters for the event tree in Figure 4.1 by ω. If the event tree in Figure 4.1 is evaluated with parameters ω, let the resulting values of AFP and EAD be denoted AFP(ω) and EAD(ω). One can then ask what statistics should be calculated for the purposes of planning and project evaluation. A reasonable and simple procedure would be for planners and engineers to select their best estimate of ω, denoted here as ω_{best}, and employ the value of AFP and EAD calculated with that best estimate:

$$AFP(\omega_{best}) \text{ and } EAD(\omega_{best}).$$

This is what is done in many studies. It is generally satisfactory when model parameter uncertainty is small.

Alternatively, if a probabilistic description has been developed to describe the likelihood of different values of ω, a different method could be employed. Just as EAD(ω) is obtained by averaging over the probability distribution for annual floods, one could average the values EAD(ω) over the probability distribution for ω. The resulting descriptions of average flood risk and average economic losses are the average annual failure probability (denoted Avg[AFP]), and the average expected annual damages (denoted Avg[EAD]), where

$$Avg[AFP] = E_{over\,\omega}\{AFP(\omega)\}$$
$$Avg[EAD] = E_{over\,\omega}\{EAD(\omega)\}$$

and where E denotes expectation over the indicated variable. The choice between AFP(ω_{best}) and Avg[AFP] and between EAD(ω_{best}) and Avg[EAD] reflects a philosophical choice in planning. The choice should also reflect how well planners believe the available distribution for ω has been specified. Even if ω_{best} is simply the average value of ω, because of the nonlinear relationship between a probability distribution's parameters and exceedance probabilities, there will generally be a difference between the two descriptions of AFP and EAD.

Whether one uses average values of AFP and EAD or uses values of AFP and EAD computed using ω_{best}, those single values should be augmented with a description of their uncertainty, which results from specification errors in the model parameters ω. One of the major contributions of risk and uncertainty analyses is the quantification of specification errors and other uncertainties, the evaluation of the resultant uncertainty in predictions and estimates of benefits and costs, and quantification of the value of additional information (Morgan and Henrion, 1990; NRC, 1994, pp. 160-61; 184-85).

A description of the uncertainty in AFP and EAD can be computed by a Monte Carlo sampling procedure using the distribution of ω to determine the distributions of AFP(ω) and EAD(ω), or just the standard deviations. The distributions of AFP and EAD can be used to determine the distribution of the benefit-cost ratio (BCR), the probability that the national development objective is less than zero for a particular project alternative, or the probability that BCR is less than one. (Such calculations and ideas were illustrated in USACE, 1992b and NRC, 1994, p. 180.)

ESTIMATION OF FLOOD DAMAGE INCORPORATING HYDROLOGIC UNCERTAINTY

Congress asked the committee to consider the issue of expected probability. An expected probability adjustment to flood frequency curves is a method that has been employed by USACE to incorporate hydrologic uncertainty into flood risk assessments (Beard, 1960, 1978). With the new risk-based planning methodology being employed by USACE that correction is no longer made explicitly. However, by including hydrologic uncertainty in its Monte Carlo evaluation of expected flood damages, USACE has implicitly introduced hydrologic parameter uncertainty (frequency-curve parameter-specification error) into the flood risk and expected damage calculations. Adding discharge-quantile uncertainty into the Monte Carlo evaluation of flood damages corresponds to what has been called an "expected damages" approach (Arnell, 1989), as opposed to the "expected probability" correction that Beard (1960, 1978, 1990) advocated.

In the framework described above, the choice would be between the use of planning criteria such as AFP(ω_{best}) and EAD(ω_{best}), which use "best" available estimates of the parameters, and Avg[AFP] and Avg[EAD], which incorporate parameter uncertainty into the estimates of those planning criteria. Reasonable arguments suggest that use of Avg[AFP] and Avg[EAD] should be entirely appropriate. However, this is true only if uncertainty in the various parameters is described well. The analysis in this section shows that use of classical statistical ideas to incorporate uncertainty into the evaluation of Avg[AFP] results in a biased exceedance probability estimator and biased estimators of flood damages. Thus the use of the classical approach is not recommended.

An Example to Consider the Estimation of Flood Damages

Consider the statistical performance of flood damage estimators for a relatively simple situation to illustrate the consequences of including hydrologic and other specification errors and uncertainties in the evaluation of flood damages. In particular, consider flood peaks Q whose logarithms $X = \log(Q)$ are normally distributed with mean μ and standard deviation σ. This corresponds to a log normal distribution for flood peaks Q, which is a special case of the Water Resources Council log-Pearson type 3 distribution, with a log-space skewness of zero (IACWD, 1982).

Let M be the sample mean and S the sample standard deviation of the available systematic record of the logarithms $\{X_i\}$ of gauged flood flows $\{Q_i\}$. Let q be a discharge of interest, possibly the discharge necessary to inundate significant buildings or overtop an existing levee designed to protect property. Then, following the Water Resources Council's (WRC's) Bulletin-17B, an estimator of the true but unknown cumulative (non-exceedance) probability $p(q)$ associated with q is (IACWD, 1982)

$$\hat{p}(q) = \Phi\{(\log(q) - M)/S\}$$

where Φ is the cumulative distribution function of the standard normal distribution.

If the flood damages corresponding to levee failure or overtopping are zero for flood peaks less than the fixed flow q, and have constant value D for flood peaks greater than or equal to q, then the true but unknown expected damages ED are

$$ED = E\{\text{Damages}\} = DP[Q \geq q] = D[1 - \Phi\{\log(q) - \mu)/\sigma\}]$$

where μ and σ are the true but unknown mean and standard derivations of flood flow logarithms X_i.

For a model with fixed damages when a levee fails or is overtopped, the conventional estimator of flood damages is

$$\text{Estimator-of-}ED = D[1 - \hat{p}(q)] = D[1 - \Phi\{(\log(q) - M)/S\}]$$

This is a very simple description of the nature of flood damages, and is used in the investigation below. In a river system like the American, the flood stage in protected areas after a levee fails may approach that in the river, and the water level beyond the levees could continue to rise if the river continues to rise. The simpler model of damages employed above illustrates the significant relationship between the estimated probability that various stages are exceeded and the estimated damages that would be computed.

The analysis above was based on the assumption that q is fixed and independent of the sample statistics M and S. Beard (1990) was critical of this assumption. Indeed, the value of q may be affected by floods that have occurred: when siting a building, the owner might have knowledge of the flood peaks that had occurred before that time. In the case of the American River, the historical levee system and Folsom Dam have provided protection from natural flows for the last three decades, so there should be little relationship between the events in the flood record and recent construction. Of course, older construction could not have anticipated the magnitude of subsequent flood peaks. Thus it is reasonable to ignore the possible interaction between the magnitude of events in the flood record and the location and value of property in the floodplain.

Beard proposed another model for flood damages that would place the property at risk at a stage corresponding to a flow $M + tS$ for some fixed scalar t (Beard, 1990). Thus the location of valuable property would be determined completely by the sample mean M and sample standard deviation S of the logarithms of the flood record that would be available when a study was performed. This is clearly an impossibility for older property and represents for newer property unusual social responsiveness to revealed flood hazard. In general, it is not a credible basis for a flood damage model.

Analysis Without Hydrologic Uncertainty

From both risk and economic loss points of view, the accuracy of any estimator of the cumulative probability $p(q)$ associated with a fixed critical flow q are of great importance to the accuracy of the calculation of expected damages. For the simple damage model discussed above, the expected damages are simply $D[1 - p(q)]$. Unfortunately, many estimators are biased, which means that their expected values (or average values over many samples) do not equal the target value. The difference between the expected value and the target value is called the "bias" of the estimator. A Monte Carlo experiment was conducted to evaluate the expected value of $[1 - \hat{p}(q)]$ for flood records of length $n = 10, 25, 50, 100$, and 200, with q ranging from the 10 percent to the 0.1 percent chance exceedance events, denoted $q_{0.1}$ and $q_{0.001}$, respectively. The results in Table 4.1 are based on 1000 generated samples yielding different values of M and S. In this instance, the results do not depend on the assumed parameters for the normal distribution describing the logarithms of the flood flows.

One can see that there is relatively little bias in the estimators of the exceedance probabilities of $q_{0.1}$ up to $q_{0.01}$ when $n \geq 50$. For $q_{0.002}$ and $q_{0.001}$ the bias is more severe, particularly for small n.

Analysis Incorporating Hydrologic Uncertainty

The new USACE risk analysis procedure proposes to include hydrologic

TABLE 4.1 Average Value of Estimated Exceedance Probability $[1 - \hat{p}(q)]$ for Specified Critical Flow q Using Sample Mean and Variance from Normal Samples of Size n Without Adjustment for Hydrologic Uncertainty

n	Critical Flow					
	$q_{0.1}$	$q_{0.04}$	$q_{0.02}$	$q_{0.01}$	$q_{0.002}$	$q_{0.001}$
10	0.11	0.048	0.028	0.016	0.0055	0.0035
25	0.10	0.044	0.023	0.013	0.0034	0.0020
50	0.10	0.042	0.022	0.011	0.0027	0.0015
100	0.10	0.041	0.021	0.011	0.0024	0.0012
200	0.10	0.041	0.021	0.010	0.0022	0.0011

uncertainty associated with the sample estimates M and S into calculation of expected annual benefits to reflect these model-parameter specification errors. If this is done correctly without approximations, it is equivalent to the expected probability adjustment that was previously employed. (Arnell (1989) analyzed both cases.) With the expected probability adjustment, the probability distribution describing the distribution of floods X is a Student t distribution with location M, scale parameter $S(1 + 1/n)^{0.5}$, and degrees of freedom n-1. This expected probability model has been employed because the ratio

$$(X - M) / [(1 + 1/n)^{0.5} S]$$

has a Student t distribution when X, M, and S are all considered to be random (Moran, 1957; Beard, 1960; Stedinger, 1983a). That analysis is also the basis of the expected probability adjustment for the log-Pearson type 3 distribution described in Bulletin 17B (IACWD, 1982, Appendix 14, p. 14).

Practically, the expected-probability adjustment yields an estimated flood frequency distribution defined by the quantile estimators

$$X_p = M + (1 + 1/n)^{0.5} t_{p,n-1} S$$

where $t_{p,n-1}$ is the pth quantile of the Student t distribution with n-1 degrees of freedom. For clarity, this estimate is called the "expected-probability quantile estimator."

There is a temptation to assume that because the expected-probability quantile estimator X_p provides a good estimate of the design flood that is exceeded with the target probability, then the corresponding frequency curve would be an appropriate relationship for estimating the probability that various fixed flow values are exceeded. However, because of the nonlinear transformations involved, the inverse of the expected-probability quantile estimator is not particularly good for estimating exceedance probabilities of fixed flood flow values.

In particular, to use the expected-probability quantile estimator X_p to evaluate the risk that a specified flow q is exceeded, one solves

$$\log(q) = M + tS (1 + 1/n)^{0.5}$$

for the corresponding t value

$$t = [\log(q) - M] / [S (1 + 1/n)^{0.5}]$$

and then looks up the corresponding cumulative probability in the tables of the Student t distribution. That calculation yields the probability estimator

$$\hat{p}^{\#}(q) = F\{[\log(q) - M] / [S (1 + 1/n)^{0.5}]\}$$

where F is the Student t probability distribution function with $(n - 1)$ degrees of freedom. For clarity, this estimate is called the "expected-probability probability estimator."

The Monte Carlo experiment described in Table 4.1 was repeated to evaluate the expected value of the expected-probability probability estimator $[1 - \hat{p}^{\#}(q)]$ for the same cases. Again, the value of q is fixed, as are the parameters of the normal distribution describing the logarithms of the flood flows. The results are reported in Table 4.2.

For every value of the sample mean and variance, the expected probability adjustment increases the estimated exceedance probability associated with the critical flow q. The estimators in Table 4.1 that ignored hydrologic uncertainty had some upward bias. The expected probability adjustment makes that bias worse in every case considered. Still, there is relatively little bias in the estimators of the exceedance probabilities of $q_{0.1}$ up to $q_{0.01}$ when $n \geq 100$. For the larger thresholds $q_{0.002}$ and $q_{0.001}$, the bias is severe, particularly for $n \leq 100$.

TABLE 4.2 Average Value of Estimated Exceedance Probability $[1 - \hat{p}(q)]$ for Specified Critical Flow q Using Sample Mean and Variance from Normal Samples of Size n with Expected Probability Adjustment Reflecting Hydrologic Uncertainty

| n | Critical Flow | | | | | |
	$q_{0.1}$	$q_{0.04}$	$q_{0.02}$	$q_{0.01}$	$q_{0.002}$	$q_{0.001}$
10	0.13	0.069	0.046	0.031	0.0148	0.0111
25	0.11	0.053	0.031	0.018	0.0063	0.0041
50	0.11	0.047	0.025	0.014	0.0040	0.0023
100	0.10	0.043	0.023	0.012	0.0029	0.0016
200	0.10	0.042	0.021	0.011	0.0025	0.0013

The natural (or conventional) estimator $1 - \hat{p}(q)$ of the exceedance probability of a levee or other flood control structures is upwardly biased. An expected probability adjustment makes that bias larger and thus does not seem to be advisable if Avg[AFP] is to be used as a decision criterion.

Others have debated the issue of bias in calculations of expected flood damages (Hardison and Jennings, 1972, 1973; Thomas, 1976; Doran and Irish, 1980). Gould (1973) and Stedinger (1983a) noted that the expected probability method was likely to increase the bias in expected flood damage. Arnell (1989) provided a clear analysis of expected annual flood damage estimates obtained with (1) the conventional estimator without a correction for hydrologic uncertainty, (2) the expected probability estimator described above, and (3) an expected damage method.

The expected damage method computes for every probability level p the expected damages associated with floods with cumulative probability p given the uncertainty in the discharge associated with p; it then integrates those "expected damages" over p. This is what the new USACE risk-based planning method does using Monte Carlo simulation (D. Ford, consultant, personal communication, January 19, 1994).

Arnell (1989) provided results such as those in Tables 4.1 and 4.2 in addition to considering two different nonlinear damage functions that begin at thresholds with exceedance probabilities of 20, 4, and 1 percent. The results obtained with those damage functions are like the results above, except that the biases are generally larger because more of the damages occur with flood flows substantially greater than the threshold flows. He concluded that,

> all methods overestimate expected annual damages, particularly when damage commences in infrequent events, but the conventional method is least biased. . . . Although the degree of difference varies with damage function, the results clearly show that the use of either expected probabilities or the "expected damage" method would produce very biased estimates of expected annual damages.

Explanation of Bias

It is useful to understand why the expected probability adjustment, which has a legitimate theoretical motivation, results in such a biased estimator of the probability that existing structures at fixed locations would be flooded. Consider its origin (Moran, 1957; Beard, 1960). Let $X = \log(Q)$ be the normally distributed logarithm of a possible future flood flow Q, where M and S are the sample estimators of the mean and standard deviation of X based on possible historical samples of size n. Q, X, M, and S are all considered to be random variables. Then for random X and M, the difference $(X - M)$ is normally distributed with mean zero and variance

$$\text{Var } [X - M] = \sigma^2(1 + 1/n)$$

As a result, the random variable defined as

$$T = (1 + 1/n)^{-0.5}(X - M)/S$$

has a Student t distribution with $n - 1$ degrees of freedom.

Let $t_{p,n-1}$ be the quantile of the Student t distribution with cumulative probability p. Then a possible future flood X will exceed the random quantity

$$\tilde{X}_p = M + t_{p,n-1}S(1 + 1/n)^{0.5}$$

on average (over the distributions of X, M, and S) with probability $(1 - p)$. This means that if many projects are built nationwide, using design flows $\tilde{X}_p$ based on calculated sample means M and standard deviations S, the nationwide annual failure rate will be equal to the target probability $(1 - p)$.

Here $\tilde{X}_p$ can be viewed as an estimator of $\mu + z_p\sigma$. Why not use

$$\hat{X}_p = M + z_p S$$

instead? Stedinger (1983a) showed that on average the probability that a future (and therefore random) X exceeds the random quantity $\hat{X}_p$ is greater than $(1 - p)$, sometimes substantially in small samples.

To understand these issues, observe that $\hat{X}_p$ is a nearly unbiased estimator of the quantile of interest $\mu + z_p\sigma$, and X does exceed $\mu + z_p \sigma$ with probability $(1 - p)$. When $(1 - p)$ is small, say 1 percent, consider what happens when the estimator $\hat{X}_p$ happens to be too large: then X exceeds this random quantile estimator with a probability less than 1 percent, perhaps 0.7 percent. On the other hand, consider what happens when $\hat{X}_p$ is too small: then X exceeds this random quantile estimator with a probability that can be much greater than 1 percent, perhaps 1.5 or 2 percent. This asymmetry results from the curvature of the cumulative probability function of X for x values corresponding to p near 1. The result is that the random quantile estimator $\hat{X}_p$ is exceeded too frequently on average (over the M and S distributions). The expected probability correction yields an estimator $\tilde{X}_p$ of p for which the exceedance probability averaged over the distributions of M and S is indeed $1 - p$.

This, however, is different from the situation where one needs to evaluate the expected damages for existing structures at fixed locations or river stages. Then to evaluate the probability of flooding for any predetermined and fixed stage x, an estimate of $(x-\mu)/\sigma$ is required. The conventional estimator $(x-M)/S$ is itself nearly unbiased, but again because of the curvature of the cumulative probability function, values of $(x-M)/S$ that are too small assign relatively more exceedance probability to the fixed flow x than values of $(x-M)/S$, which are too big and correspond to relatively smaller exceedance probabilities. Table 4.1 shows that use of

$$\hat{p} = \Phi\{(x - M)/S\}$$

yields exceedance probabilities for the fixed flow x that on average are too large. Table 4.2 shows that use of the expected probability estimators results in an even greater bias. This is due to use of the Student t distribution and the scale factor $(1 + 1/n)^{0.5}$.

To eliminate these biases, one might propose a downward adjustment of the probability curve (an expected-damage bias-adjustment correction for damage sites at fixed locations). The committee does not propose such an adjustment. The conventional estimator without adjustment for hydrologic uncertainty can be thought of as a compromise.

For designing a nationwide system of small structures, such as highway culverts, to meet mandated failure-rate criterion, the expected-probability adjustment can be appropriate. However, when evaluating the expected damages associated with existing structures and population centers at fixed locations and stages, the conventional estimator without a correction for expected probabilities or hydrologic uncertainty is the better choice.

The Bayesian Viewpoint

The problem of hydrologic uncertainty, to the extent it is due to records of limited length, represents a classic statistical sampling problem. The statistical literature contains two methods for representing such uncertainty. Confidence intervals are a means of expressing uncertainty in terms of intervals that in repeated sampling will bracket the true value of a parameter with a specified frequency, called the confidence of the interval. They are the most commonly used method of describing sampling uncertainty.

The second method employs Bayesian statistics (sometimes called inverse probability). Bayesian procedures describe the possible values of a parameter by a probability distribution that represents an engineer's or statistician's degree of belief as to the likelihood that the parameter has different values. Such distributions generally depend upon the available data as well as the engineer's or statistician's prior beliefs and other sources of information such as regional hydrologic experience (Benjamin and Cornell, 1970). Many examples of Bayesian procedures in flood frequency analysis are available (e.g., Davis et al., 1972; Vicens et al., 1975; Wood, 1978; Stedinger, 1983a; Bernier, 1987). There are also empirical Bayesian methods that explicitly use regional information to eliminate use of subjective distributions to describe prior beliefs and regional experience (Kuczera, 1982).

To illustrate the importance of nonsample information, consider a hydrologist's estimate of a probability distribution (which in a Bayesian analysis is called the posterior probability distribution) for the probability that a levee is overtopped based on (1) 20 years of data without any levee overtopping events and (2) a reasonable understanding of the hydrology of the basin. While the sample data may have convinced the hydrologist that a levee overtopping event is unlikely, there would also surely be some risk of such an event. Thus a reasonable estimate of the probability of levee overtopping would not be zero but should reflect a broader sense of what is likely and physically reasonable.

There are important conceptual and practical differences between the classical statistical approach and a Bayesian analysis. In particular, a proper Bayesian analysis employs an informative prior distribution for the unknown parameters (such as the probability of levee overtopping in the example above). That information may result in the posterior distribution having a smaller or a larger mean and variance than the sample moments M and S^2. As a result, a Bayesian analysis will provide, on average across basins, an unbiased estimator of flood damages (Stedinger, 1983a); thus the introduction of a legitimate prior distribution is very important. However, this is different from an estimator being unbiased at each site. In cases where the available data overwhelm a proper informative prior distribution, parameter uncertainty is likely to be relatively unimportant, and it should not matter whether classical or Bayesian methods are employed.

It is interesting that the distribution obtained for X with an expected probability adjustment is equivalent to a Bayesian posterior distribution using a so-called noninformative prior (Stedinger, 1983a). So why is there a problem with an expected probability adjustment if it is equivalent to a Bayesian estimator? The problem is that the expected probability estimator results from use of a noninformative Bayesian prior, which always inflates the mean and variance of the Bayesian posterior flood distribution. Thus it assigns an infinite prior mean and prior variance to flood flows. As a result, it is not surprising that the expected probability adjustment always yields an upward biased estimator of flood risk and damages.

Recommendation

To avoid the problem of bias in estimating expected annual damages, it seems most appropriate that the economic assessment and descriptions of the probability of flooding be based on best estimates of the parameters of models, $AFP(\omega_{best})$ and $EAD(\omega_{best})$. These two statistics still involve calculation of the expectation over significant processes contributing to flood risk and variability in system operation, perhaps as illustrated in Figure 4.1. The alternative would be for USACE to adopt a correct Bayesian analysis of flood risk uncertainty with a proper informative prior based on regional hydrologic information as well. Given the lack of precedence, and the need for uniform and accepted procedures for the selection of prior distributions to describe uncertainty in the parameters of important models, use of proper Bayesian procedures does not appear feasible at this time.

The use of $AFP(\omega_{best})$ and $EAD(\omega_{best})$ as the primary criteria to summarize the most likely performance of a project has other advantages as well. They will not be dependent on the descriptions of uncertainty employed to describe hydraulic and economic models. Thus they will be conceptually easier to understand. Moreover, the descriptions of uncertainties used to describe many of these processes will be fairly subjective and not particularly well determined. While it is often a challenge to determine the best estimators of each of the processes in Figure 4.1, it is much more difficult to describe well the uncertainty in those parameters. Use of $AFP(\omega_{best})$ and $EAD(\omega_{best})$ will separate the description of the likely operation of the system from problems related to the description of uncertainties.

These best estimates, $AFP(\omega_{best})$ and $EAD\omega_{best})$ should still be supplemented by descriptions of their accuracy. In particular, possible values of AFP and EAD could be generated by Monte Carlo simulation procedures to illustrate the uncertainty in these performance criteria. Similarly, the uncertainty in AFP and EAD could be described by their standard error or particular quantile ranges, while the impact of that uncertainty could be illustrated by the probability that the national economic development objective is negative. Those descriptions of uncertainty in AFP and EAD and its impact would depend on the selected representations of uncertainty in hydrologic, hydraulic, and economic parameters, and would allow agency planners and the public to assess the accuracy of $AFP(\omega_{best})$ and $EAD(\omega_{best})$. NRC (1994, pp. 184-85) made similar recommendations for health risk analyses and also recommended that risk assessors carefully explain qualitatively the basis for such numbers so as to minimize public misunderstanding (NRC, 1994, p. 13).

This proposal appears to be consistent with the requirement in EC 1105-2-205 (1994, p. 4) that "the estimate of NED benefits will be reported both as a single expected value and on a probabilistic basis (value of the benefit and its associated probability), for each planning alternative." Table B-3 in EC 1105-2-205 (USACE, 1994) illustrated the presentation of economic and risk criteria for different project alternatives (corresponding to levee heights) and included with-project residual damages (equivalent to Avg[EAD]) and the median probability of exceedance (equivalent to $AFP(\omega_{best})$). The table also reports the simulated stage-exceedance probability (equivalent to Avg[AFP]). $AFP(\omega_{best})$ and $EAD(\omega_{best})$ include expectations over the distributions of annual maximum flood events and perhaps also reservoir operation and levee failure stage. They would also be based on the expected value of damages for each stage in a flood damage zone. EC 1105-2-205 (in its Appendix A) described the derivation of such an expected value and the associated uncertainty.

METRICS FOR PROJECT PERFORMANCE EVALUATION

Fundamental questions in flood protection project evaluation are,

- What is the probability that target areas will be flooded?
- Do economic parameters justify proposed projects?

and perhaps,
- How reliable is the economic analysis of alternative projects?

Thus a thorough risk-based flood-protection analysis should calculate the following values for each project alternative:

1. The best estimate of the annual probability of flooding for target locations, called the annual failure probability (AFP).
2. The expected economic benefits and costs for each project, based on the expected annual damages (EAD).
3. Measures of the uncertainty, lack of accuracy, or likely specification errors associated with (1) and hopefully also (2).
4. Measures of the reliability of system performance that contribute to an engineer's and the public's understanding of system dynamics and how individual components of the system are likely to perform.

While project selection is for the most part determined by an economic evaluation of the alternatives, the best estimate of the risk of flooding at target locations is perhaps of most interest to the public and many public officials. They are interested in risk of flooding both without any project and with various alternative projects. For this reason, this performance index has been listed first.

The justification for most projects is ultimately economic, though environmental, social, and equity impacts should not be neglected. Thus the second performance criterion is the economic efficiency or national economic development (NED) objective, which depends on the expected annual damages associated with different alternatives.

Because of the importance of the economic evaluation of a project, and the inherent uncertainty in the performance of flood protection projects, it is important to evaluate the uncertainty in the economic performance of alternative projects. This point was made clearly in USACE (1992a); USACE (1992b) provided an extended example of the calculation of the uncertainty in estimates of the national economic development objective and the benefit-cost ratio (BCR). Economic uncertainty was illustrated by the probability that the BCR is actually less than one, the mean and coefficient of variation of the BCR, the distribution of net benefits, and the distribution of the BCR (USACE, 1992b). All of these are reasonable approaches for illustrating the impact of uncertainties that affect the economic performance of a project.

The uncertainty in the economic performance of a project depends largely on the uncertainty associated with flood flow frequency distributions, hydraulic relationships, specification of when levee failure occurs, and the economic value of property and the damages it might sustain. EC 1105-2-205 illustrated the impact of structural and content values in the evaluation of the uncertainty of economic measures of project performance. Following Davis (1991) in the application of the USACE risk and uncertainty methodology, the Sacramento District also proposed to calculate various reliability indices to help explain project performance. Clearly, the value of such indices is less important than either the overall risk of flooding faced by residents of the floodplain or the relative economic attractiveness of alternative projects. Moreover, the attractiveness of alternative plans should be judged primarily on risk of flooding and economic efficiency rather than on whether some internal measure of system reliability meets an arbitrary standard. These issues are discussed in greater detail later in this chapter.

USACE RISK-BASED PROCEDURES

The sections above discuss a general formulation and structure of risk and uncertainty analyses for flood protection project evaluation, and metrics for project evaluation. The committee's task included looking carefully at the risk and uncertainty methodology used by the Sacramento District. The sections below first review the general approach USACE has adopted and then focus on the implementation of that philosophy in the American River basin.

The New Methodology

USACE (1994) has observed that "risk and uncertainty are intrinsic in water resources planning and design" (EC 1105-2-205, 4(a)). In the past, USACE first developed its best estimate of the most likely values of "key variables" for the evaluation and design of flood damage reduction projects. Then sensitivity analysis was used as the primary tool to investigate the importance of uncertainty in planning parameters. However, this approach fails to integrate sources of uncertainty, their interaction, and their relative likelihoods (Moser, 1994). The new USACE risk-based procedures quantify the risks and uncertainties in various parameters and components of the planning process and the design of facilities. USACE has described the new risk-based analysis framework as "an approach to evaluation and decision making that explicitly, and to the extent practical, analytically incorporates considerations of risk and uncertainty."

USACE has decided that the traditional "level of protection" will no longer be used in describing project performance, and levee freeboard will be replaced with a probabilistic description of levee performance (EC 1105-2-205). Previously, levee height determination was caught between the designer's view that it should ensure that a project can reliably pass the design flood, and the economist's view that freeboard should be economically justified. The new risk-based analysis will allow USACE analysis to address the economic and reliability tradeoffs associated with levee freeboard (Davis, 1991). EC 1105-2-205 indicated that,

"Risk-based analysis enables risk issues and uncertainty in critical data and information to be explicitly included in project formulation and evaluation" (EC 1105-2-205, p. A-1).

The new risk-based decisionmaking procedures combine traditional hydrologic risk with

- uncertainty in parameters describing hydrologic risk (hydrologic uncertainty),
- variability and uncertainty in stage-discharge (hydraulic) relationships,
- levee performance variability,
- variability and uncertainty in stage-damage (economic) relationships, and
- variability in other operating assumptions.

Some of these uncertainties arise because of limited hydrologic records, while others reflect limited data and errors in measurements of channel geometry, roughness and slope, or the range and character of economic activities. USACE (EC 1105-2-205, 1994, p. A-3) observed that "The proposed strategy is similar to present practice but differs in that uncertainty is explicitly quantified and integrated into the analysis."

In addition, USACE risk-based analyses will initially concentrate on including uncertainty in the following key variables (Moser, 1994):

Economic	**Hydrologic**	**Hydraulic**
Structure, first-floor elevation	Discharge associated with exceedance frequency	Conveyance roughness
Structural values		Cross-section geometry
Content values		

Figure 4.2 (from Davis, 1991; also EC 1105-2-205, p. A-2) illustrates possible uncertainties in discharge-frequency, stage-discharge and damage-stage relationships. Table 4.3 summarizes the sources of risk, variability, and uncertainty that can be considered in a risk and uncertainty-based planning methodology.

Reasons given to support this evolution in the USACE planning methodology include

1. The inherent uncertainty in planning.
2. That hydrologic and hydraulic engineering have advanced sufficiently to reduce the need to design for unquantified uncertainties. As a result the degree of certainty in performance (engineering reliability) can be quantified and advantageously used in project selection.
3. The broader range of risk, costs, reliability, and associated trade-offs that it will allow the planning process to address (Moser, 1994, pp. 1 to 2).

Flood Risk

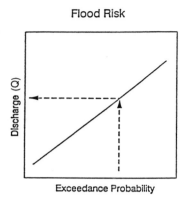

Uncertainty in Discharge

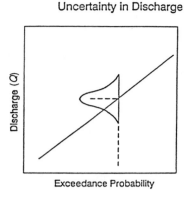

Uncertainty in Stage

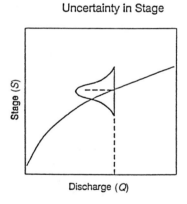

Uncertainty in Damage

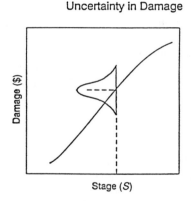

FIGURE 4.2 Uncertainty in discharge, stage, and damage.

USACE indicated that its risk-based planning procedures should allow examination of the total effect of risks and uncertainties on design values and economic viability. Thus better decisions can be made on the trade-offs between risks and costs. Increasingly, USACE is confronted with severe budget constraints, new customer cost-sharing requirements, and concern among its customers and the public with project performance and reliability. System performance and planning uncertainties now need to be addressed more explicitly as part of the assessment of water resource investments. The new risk-based procedures are

TABLE 4.3 Sources of Risk, Variability, and Uncertainty That Can Be Considered in a Risk and Uncertainty-Based Planning Methodology

1. Hydrologic risk: Discharge Q associated with exceedance probability p.
2. Hydrologic uncertainty: Variability in estimators of moments of the Q-distribution and the accuracy of derived frequency curves.
3. Flood stage: stage S corresponding to discharge Q is not perfectly determined owing to variability and imperfect knowledge of channel geometry, roughness, flow regime, bed form, flow debris, and inexact analytical techniques.
4. Levee performance: Stage L at which levee fails is uncertain owing to lack of understanding of internal structure, and possibility of surface erosion, piping problems, underseepage, slides within the levee embankment, and foundation soil weaknesses.
5. Flood damage uncertainty: Damage D on floodplain associated with river stage S is uncertain owing to mix of structures, elevations, and structural and content damage potential, which determines damage distribution about stage-damage curve. EC 1105-2-205 (p. A-3) indicated that damage uncertainty can describe uncertainty in the extent of the physical damage or in the "cost data."

presented with the expectation that more explicit consideration of risk and uncertainty should improve USACE investment decisions and the planning process (EC 1105-2-205, 4(d)).

Explicitly introducing hydraulic and levee-performance variability into the analysis should improve estimates of the true overall risk of levee system failure, as well as identification of the critical processes most likely to result in failure. For example, one can ask how the risk of flooding would change as a result of rigorous inspection and some structural improvements in levees (without actually attempting to raise their crest), as opposed to developing increased flood control storage in reservoirs, which would address the issue of flood control more explicitly.

On a philosophical basis the USACE proposal is the logical and appropriate next step in the evolution from the previous flood-damage-reduction evaluation methodology. USACE is to be commended on beginning the development of this planning capability. However, the new procedures will not necessarily be easy to implement. Safety factors are often an easy way to avoid messy issues. When they are replaced by probability distributions that are actually used to describe project performance, much more attention needs to be paid to some critical uncertainties. Models will be needed to represent when levees actually fail as a function of stage, levee characteristics, and levee length. The need for these functions is new, and they may be relatively important. Likewise, variability in hydraulic calculations defining stage-discharge relationships will also be needed; such calculations are an extension of traditional hydraulic sensitivity analyses. USACE recognizes these problems and has embarked on a vigorous research effort (EC 1105-2-205, Appendix C).

The ability of the new planning procedure to better represent flood risk, system design, and operating trade-offs depends on how well many modeling issues are handled in each application of the procedures. Providing general guidance and training for district engineers on procedures and data bases would strengthen the engineering judgments and modeling assumptions they make. Instituting report review at the district and national level should also improve consistency and accuracy. Using USACE research organizations as centers of excellence would help to develop and disseminate the needed expertise and experience.

The committee realizes that the increased complexity of the new risk and uncertainty-based analyses may lead to less reliable estimates of flood risk such as Avg[AFP], at least until the method is better understood by those using it. For example, it will be hard to capture the effectiveness of flood-fighting efforts, and the feedbacks between system weaknesses and variations in reservoir operations to avoid failure. Nevertheless, ultimately the new methodology can provide a more accurate estimate of the true residual flood risk associated with a project and the uncertainty in average estimates of performance criteria due to unavoidable model specification errors. Engineers also should be careful not to mix the new estimates of the chance of flooding, such as Avg[AFP], with the old "level of protection." Such confusion is a major problem with the 1994 Alternatives Report (USACE, Sacramento District, 1994a). The two approaches have different sets of assumptions.

Description of Risk and Uncertainty in the New Methodology

This section provides a brief description of the treatment of risk and uncertainty in the new USACE risk-based planning methodology. The structure of this section follows Table 4.3.

Hydrologic Risk

More effort and work have gone into capturing and describing hydrologic risk than have gone into the other issues. As a result, it is much easier to criticize the procedures in light of the many flood frequency procedures that have been proposed. USACE employs WRC Bulletin 17B procedures (IACWD, 1982), which are recommended federal guidelines, and uses a log-Pearson type 3 (LP3) distribution to describe the frequency relationship. Issues associated with that procedure and recently developed alternatives are discussed elsewhere (Thomas, 1985; Potter, 1987; Cunnane, 1988; Potter and Lettenmaier, 1990; Stedinger et al., 1993).

Hydrologic Uncertainty

Hydrologic uncertainty is simpler to deal with than other sources of uncertainty, when the analysis is based on a stationary gauged record. For the most part, hydrologic uncertainty in estimators of the parameters is determined by the limited length of the flood series used. In that sense, the uncertainty is objective and is described by standard statistical sampling theory (IACWD, 1982; Stedinger, 1983b; Chow et al., 1988; Chowdhury and Stedinger, 1991).

When a flood record must be corrected for development, storage, or channel changes, then the length of record is still likely to be the primary determinant of hydrologic uncertainty, though subjective assessments of the quality of any adjustments to measured flows are also important. Possible nonstationarity due to subtle shifts in climate and storm paths is difficult to detect and document, but is sometimes a concern. If regional relationships are used to develop flood curves, then the corresponding estimates of prediction error should be employed (Tasker and Stedinger, 1989).

The proposed analysis for gauged sites bases its description of hydrologic uncertainty upon the confidence interval calculation procedure in Bulletin 17B (IACWD, 1982), which contains procedures that federal agencies agreed to employ in the mid- 1970s. The Bulletin 17B procedure for calculating confidence intervals employs the assumption that the coefficient of skewness of the logarithms of the floods is correctly specified, independent of the data (Stedinger, 1983b). In fact, the actual coefficient of skewness employed is generally a weighted average of the at-site sample skewness and a regional or generalized skewness estimator (IACWD, 1982). Clearly, the weighted skewness estimators incorporate estimation error because of sampling error in the at-site skewness estimators and also the regional skewness estimators (McCuen, 1979; Tasker and Stedinger, 1986). As a result the calculated intervals with the Bulletin 17B procedure are too small. Formulas that incorporate variability in weighted skewness estimators are available (Chowdhury and Stedinger, 1991; Stedinger et al., 1993).

As noted above, the problem of hydrologic uncertainty, to the extent it is due to records of limited length, represents a classic statistical sampling problem. The two approaches in the statistical literature for representing such uncertainty are traditional confidence intervals, which are interval estimators that contain an unknown but fixed parameter with a specified frequency, and Bayesian inference, which describes the uncertainty in unknown parameters by a probability distribution. The USACE risk and uncertainty methodology employs a Monte Carlo procedure to generate alternative values of flood quantiles reflecting hydrologic uncertainty. This is used to estimate the probability of levee failure and expected annual damages. Given this approach, it would appear that USACE would need to adopt a Bayesian framework to be conceptually consistent. In a Bayesian framework hydrologic variability and uncertainty are integrated to obtain the posterior distribution for flood flows (Zellner, 1971). Descriptions of the uncertainty in flood quantiles and flood-distribution parameters by probability distributions are inconsistent with the theory supporting traditional confidence intervals because that theory depends on the parameters being fixed; only in repeated sampling does the concept of "confidence" associated with a confidence interval have meaning (Stedinger, 1983a). However, the USACE procedure employs the confidence interval procedures from Bulletin 17B to generate alternative flood quantiles associated with each generated exceedance probability p and then uses these values to compute a probability of flooding.

Flood Stage Uncertainty

At some gauged sites the uncertainty in stage estimates for a given flow is related to the accuracy of the stage-discharge rating curve and its stability. Agencies such as the U.S. Geological Survey that are responsible for the estimation and updating of the rating curve should be able to provide information about its accuracy and stability. In most cases, hydraulic models will be required to compute water surface elevation profiles based on surveys of channel capacity and in some cases perhaps also on the operation of storage facilities. At some locations the analysis may be plagued by complex hydraulics and junctions with other rivers and streams or with hydraulic control structures and weirs that divert flood flows. In such instances determination of the stage associated with a given flow may be difficult.

Development of descriptions of flood stage uncertainty at ungauged sites can be viewed as an application of sensitivity analysis. The difference is that more care will need to go into the specification of the uncertainty distribution for different parameters. First-order uncertainty analysis techniques can be used in situations with relatively small errors to derive the resulting distribution of errors in stage (Benjamin and Cornell, 1970; Burges, 1979). Mays and Tung (1992, section 5.3) illustrated the application of this method to Manning's formula for open channel flow. Kuczera (1988) discussed the accuracy of more complex rainfall-runoff calculations. USACE (EC 1105-2-205, p. A-16) has provided guidelines for the estimation of stage-discharge relations.

Levee Performance Variability

The reliability of levees will be an important component of risk-based planning studies. USACE has outlined procedures for developing these distributions based on the opinions of experts and its review of available data (Memorandum for Major Subordinate Commands and District Commands, Policy Guidance Letter No. 26, Benefit Determination Involving Existing Levees, 23 Dec. 1991).

Flood Damage Uncertainty

Flood damage estimates for residential and commercial areas are based on (1) the number of different types of buildings, (2) structural value by building type and usage, (3) contents of building by type and usage, (4) first floor elevation, (5) damage percentage as a function of flood depth, and (6) flood depths at damage locations as a function of river stage. The errors in such calculations can be estimated by considering the errors likely in each component of the analysis (EC 1105-2-205, pp. A-19 to A-39). The analysis is more difficult for the calculation of with-project stage-damage functions because one should anticipate the response of floodplain occupants to new construction and other projects that are intended to reduce flood risk.

Using the USACE Risk-Based Analysis Framework
for the American River

Information about how the risk-based procedures are being applied to the planning activities in the American River basin were provided to the committee largely in a presentation on August 11, 1994, by USACE and in letters from consultant Dr. David Ford (D. Ford, consultant, personal communications, August 23 and September 1 and 19, 1994). The risk and uncertainty analysis is summarized in Figure 4.3.

The risk and uncertainty procedures for the American River basin focus on the national economic development (NED) objective, risk of flooding, and system reliability. Expected annual damages (EAD) and expected annual failure probability (AFP) are computed by repeatedly sampling from the discharge-frequency function and the levee stage-stability probability function, as well as the error distributions for frequency, stage-discharge, and stage-damage. USACE often refers to the annual failure probability as the "annual exceedance probability." Because failure can occur with floods of different magnitudes depending on reservoir system operation and levee performance, the word "exceedance" may have lost its original meaning, which referred to a unique flood with a specified exceedance probability.

If occurrences of failure events are independent from year to year, the risk of occurrence of at least one failure in a T-year period can be computed from the annual failure probability as

$$Risk(T) = 1 - (1-\text{AFP})^T$$

American River Risk-based Simulation Analysis

The analysis procedure for the evaluation of risk and expected economic losses is based on sampling the relevant performance and uncertainty distributions, as illustrated roughly in Figure 4.2. The steps described by Ford (personal communication, September 19, 1994), after some reorganization, are repeated below. The algorithm simultaneously incorporates hydrologic risk and variability in flood operations, and hydrologic and hydraulic uncertainty. Averaging the

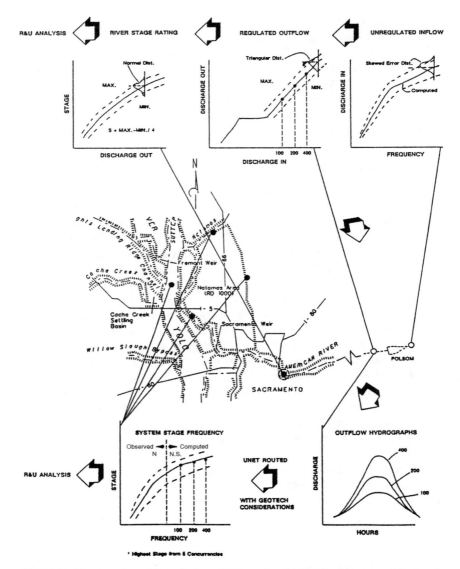

FIGURE 4.3 American River project risk-based reliability hydrology and hydraulics.
SOURCE: USACE, Sacramento District, August 11, 1994.

generated values yields Avg[AFP] and Avg [EAD]. Unfortunately, economic uncertainty is ignored in the American River study (Ford, personal communication, September 1, 1994).

Discharge-Frequency Relationship (Hydrologic Variability)

The first step of the simulation is to sample the "median" discharge-frequency function (the fitted flood-frequency curve) to obtain a nominal flood flow value Q. Bulletin 17B (IACWD, 1982) describes the procedures employed to obtain this distribution. This step corresponds to use of the frequency distribution in Figure 4.3 and in Figure 4.2, which is represented by a solid line.

Hydrologic Uncertainty

The second step incorporates hydrologic uncertainty into the analysis. Using the confidence interval procedure in Appendix 9 of Bulletin 17B (IACWD, 1982), the flood flow selected in step 2, Q, is modified to reflect possible errors in the "median" flood frequency curve. The resulting flood flow reflecting possible error in the estimation of Q is denoted $\tilde{Q}$.

This step corresponds to use of the error distribution about the frequency distribution in the upper right-hand corner in Figure 4.2 and the upper-right-hand corner in Figure 4.3. In Figure 4.2 that error distribution is represented by a bell-shaped curve.

Reservoir Operation

In the third step, the unregulated Folsom inflow must be transformed (by storage routing using the reservoir's operating rules) into a peak outflow rate for evaluation of downstream damages in the American River corridor. The Sacramento District considers variability in this transformation due to variations in initial storage, possible delays in making releases, use of a one- or two-wave model of the inflow hydrograph, outlet works operation, and spillway operation efficiency.

The worst-case, most likely, and best-case values for operational performance and decisions were analyzed to determine for various inflow levels a possible distribution of peak outflow rates. A triangular distribution for outflow was determined by assigning probabilities of 5 and 95 percent to the best- and worst-case outflows, respectively, and computing the non-exceedance probability of the most-likely outcome so as to yield a legitimate distribution function (Ford, personal communication, September 19, 1994). That distribution allows the assignment of a peak outflow reflecting possible variation in reservoir operation $\tilde{O}[\tilde{Q}_p]$ to each computed error-affected inflow with error $\tilde{Q}_p$ from step 2. This is illustrated in the middle graph at the top of Figure 4.3.

The Sacramento District will eventually need to justify the selected ranges for each of the factors considered in this step. It would also be useful for the district to provide an analysis illustrating which factors were most important in determining the variance of the outflow distribution so that attention can be focused on the key factors. In particular, if initial delays in releases are particularly important, then planners and engineers should investigate general policies, operating procedures, and warning and alert systems that might be able to reduce such delays.

Stage-discharge

In the fourth step, stages $S[\tilde{O}]$ at various locations in the American River corridor can be estimated given the Folsom outflow peak $\tilde{O}$. This is illustrated in the lower-left-hand corner of Figure 4.2 and the upper-left-hand corner of Figure 4.3.

Hydraulic Uncertainty

Hydraulic uncertainty is quantified in the fifth step. The calculated stage $S[\tilde{O}]$ is an imperfect estimator, so an error-affected stage is generated, which is denoted $\tilde{S}[\tilde{O}]$. This corresponds to the error distribution shown in the lower left-hand corner of Figure 4.2 and the upper left-hand corner of Figure 4.3.

Levee Performance

Whether or not a levee fails is determined by the height of the water in the channel, though other factors such as the duration of flooding could be important. For the sixth step USACE (Engineering Technical Letter 1110-2-328 22 March 1993) prescribes describing levee reliability for existing levees by two points: Probable Failure Point, PFP, and Probable Nonfailure Point, PNP (see Figure 4.4). As shown in Figure 4.4, there is a 15 percent probability that the levee would fail at the PNP, and the probability increases from 15 percent at the PNP stage linearly with stage until it reaches 85 percent at the PFP stage. At the PFP stage the failure probability increases discontinuously to unity. Geotechnical engineering evaluations are the basis of these two stages. The selected stages replace the use of a single stage to define when a levee would fail (with some residual freeboard included as a safety factor). However, USACE has indicated that for new levees, PNP and PFP both equal the stage of the levee crest, perhaps with an allowance for settling.

If the levee does indeed fail, then the error-affected river flow with error $\tilde{O}$ should be used to determine a new error-affected stage $\tilde{S}[\tilde{O}]$ for damage sites of interest.

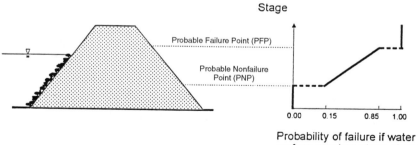

FIGURE 4.4 Failure relationship of PNP and PFP. SOURCE: USACE, letter 1110-2-328, March, 1993.

Stage-Damage Relationship

In a system without levees, the error-affected stage with error $\tilde{S}[\tilde{O}]$ from step 5 is used to determine the economic damages. This is the situation considered in the example in EC 1105-2-205 (p. A-2). In a system with levees, the situation is more complex because one must consider if the levee fails, and the flooding that would result if it does. In step 7, if a levee is overtopped or fails for other reasons, the error-affected stage, given levee failure stage $\tilde{S}[\tilde{O}]$ determines the damages.

Stage-Damage Uncertainty

EC 1105-2-205 (pp. A-19 through A-40), and USACE (1992b, pp. FC-23 through FC-33) described the origin of errors in the estimation of expected damages. Many of these are related to the limited resources available to determine the number, types, and value of structures in areas likely to be flooded. Other uncertainties relate to factors besides stage that determine the damages from flooding. These include flood duration, the presence of ice or wave action, and warning time, as well as fundamental problems in determining the costs of damage to property. These are represented in the lower-right-hand corner of Figure 4.2.

If the only statistic of interest were the expected annual damages (EAD), there would be no need to consider uncertainty in damages; it would suffice to employ with each error-affected stage $\tilde{S}[\tilde{O}]$ the average estimated damages that would result were the river to reach that stage. However, as illustrated by USACE (1992b), uncertainty in the EAD and benefit-cost ratio resulting from the uncertainty in damage estimates and the discharge-frequency relation can be substantial. It was the committee's understanding that uncertainty in estimated damages would not be part of the risk-based analysis performed in the American River study (Ford, personal communication, September 1, 1994). That is unfortunate because as a result it will not be possible to determine how uncertainty in key economic parameters might affect the ranking of flood control alternatives for the American River basin, and the overall viability of the more attractive projects.

Organization of the American River Study Analysis

There are a number of ways to organize the risk-analysis computations. The simplest procedure would be to randomly generate values of all of the variables and count the number of times the levee fails to determine the levee failure probability (which determines the AFP), and to average the resulting damages to determine the expected annual damages (EAD). Ford (personal communication, September 1, 1994) describes this procedure for calculating failure probabilities.

However, the Monte Carlo simulation described above can be simplified by introducing an analytical evaluation of levee failure, as suggested by Figure 4.1. Given a calculated error-affected stage $\tilde{S}[\tilde{O}]$, one can determine the probability of levee failure for that stage, denoted p_L. Suppose the resulting damages if the levee failed would be $D(O)$, and zero otherwise. Then the average damages associated with the generated stage $\tilde{S}[\tilde{O}]$ are just

$$p_L D(O) + (1 - p_L)0 = p_L D(O)$$

By averaging across the Monte Carlo replicates the expected damages $p_L D(\tilde{O})$ associated with each error-affected stage $\tilde{S}[\tilde{O}]$, one obtains the expected annual damages (EAD).

A similar simplification can be employed when calculating the expected annual levee failure probability. For each randomly generated stage $\tilde{S}[\tilde{O}]$, one obtains the corresponding probability of levee failure $p_L(\tilde{S}[\tilde{O}])$; the average of these probabilities is the expected annual failure probability (AFP). The committee understands that the Sacramento District employed the second of these ideas (Ford, personal communications, September 1 and 19, 1994). The two shortcuts eliminate from the Monte Carlo analysis variability due to the random generation of different levee failure stages.

Other significant simplifications are possible. As EC 1105-2-205 (p. A-3) pointed out, the problem as a whole is quite complex. As a result, analytical, or analytical-numerical evaluation of the entire problem may not be as attractive as Monte Carlo sampling schemes. However, analytical approximations could be introduced in some places to make the computations simpler and more accurate. For example, if one chooses to introduce discharge-frequency uncertainty, that uncertainty might be combined with the original "median" frequency curve to get an error-affected discharge-frequency distribution curve. That error-affected in-flow curve could then be convolved numerically with the distribution for the inflow-outflow transformation to obtain an error-affected outflow-frequency curve to eliminate that simulation step.

USACE USE OF RELIABILITY IN PROJECT PLANNING

Developers of the new USACE risk-based planning methodology have proposed use of a system reliability index as a key description of a system's ability to meet particular performance levels (Davis, 1991). Examples can be found in the 1994 Alternatives Report (USACE, Sacramento District, 1994a, Plate 12 and p. 57, Table III-1 and p. 9). In a case study illustrating the new methodology, Dotson et al. (1994) observed:

> The risk-based approach has many similarities with the present practice in that the basic data are the same. Best estimates are made of discharge/frequency curves, water surface profiles, and stage/damage relationships. The difference between the current practice and the risk-based approach is that uncertainty in technical data is quantified and explicitly included in evaluating project performance and benefits. *Using the risk-based approach, performance can be stated in terms of reliability of achieving stated goals.* Also, adjustments or additions of features to accommodate uncertainty, such as adding freeboard or levee/flood walls, are not necessary. [Italics added.]

USACE Guidelines for Use of a Reliability Index

The committee struggled to understand how this reliability index would be used and what information it conveyed. Before those issues are examined, the use of reliability in project evaluation needs to be considered. The USACE guidelines for the new risk and uncertainty analysis procedures (EC 1105-2-205) state,

> The risk-based analysis will quantify the reliability and performance of levee heights considered by explicitly incorporating the uncertainties associated with key variables. *This reliability and performance will be reported as the protection for a target percent chance exceedance flood with a specified reliability.* For example, the proposed levee project is expected to contain the one-half percent (0.5 percent) chance exceedance flood, should it occur, with a ninety percent (90 percent) reliability. This performance may also be described in terms of the percent chance of containing a specific historic flood should it occur. [Italics added.]

With these changes the directive indicates that "the concept of level of protection is no longer useful and will not be used in describing project performance."

To illustrate the concept, EC 1105-2-205 presents several examples in its Appendix B. Table B-2 from that appendix is reproduced here as Table 4.4. (An earlier version of that example appeared in Davis (1991).) For different levee

TABLE 4.4 Reliability of Design Stage for Various Events

Levee Height (ft)	Percent Chance Nonexceedance for 25-Year Event[a]	Percent Chance Nonexceedance for 50-Year Event[b]	Percent Chance Nonexceedance for 100-Year Event[c]	Percent Chance Nonexceedance for 250-Year Event[d]	Percent Chance Nonexceedance for 500-Year Event[e]
13	0.3				
14	2.6				
15	14.3	0.4			
16	36.8	2.6	0.1		
16.5	49.9	5.8	0.3		
17	62.4	11.4	0.6		
18	81.6	27.4	2.7		
19	92.2	47.3	9.2	0.3	
19.1	92.9	49.6	10.0	0.4	
20	96.8	66.6	20.5	1.2	0.1
21	98.7	80.2	35.5	3.3	0.4
21.9	99.2	88.2	48.3	6.6	0.7
22	99.2	89.0	50.3	7.2	0.8
23	99.5	94.1	63.8	13.9	1.9
24	99.7	97.0	75.0	24.0	5.2
25	99.9	98.5	85.3	40.1	13.5
25.5	99.9	99.0	89.6	48.9	19.7
26		99.3	93.1	59.3	26.8
27		99.7	97.5	76.3	44.7
28		99.9	98.9	86.9	61.1
29		99.9	99.3	93.2	74.0
30			99.6	96.6	84.8
31			99.7	98.5	92.4
32			99.9	99.5	97.2
33				99.8	99.5

NOTE: Probability distributions are based on 4,500 iterations of Latin hypercube sampling using @Risk (Palisade Corporation, 1992).

[a]Given that the median 25-year event (design stage = 16.5 feet) occurs, the probability that the event will be contained with the levee height given in the first column.

[b]Given that the median 50-year event (design stage = 19.1 feet) occurs, the probability that the event will be contained with the levee height given in the first column.

[c]Given that the median 100-year event (design stage = 21.9 feet) occurs, the probability that the event will be contained with the levee height given in the first column.

[d]Given that the median 250-year event (design stage = 25.5 feet) occurs, the probability that the event will be contained with the levee height given in the first column.

[e]Given that the median 500-year event (design stage = 27.0 feet) occurs, the probability that the event will be contained with the levee height given in the first column.

SOURCE: USACE, 1994.

stages, and different possible flood events defined by the chance of flooding, ignoring hydrologic uncertainty, the table presents the reliability of the levee (the probability it will not be overtopped). The calculated reliability includes hydrologic uncertainty, uncertainty in calculated stages given river discharge, and variability in levee failure stage. EC 1105-2-205 observes that:

> Table 2 in Appendix B shows that for a levee height of 25 feet there is a 14 percent chance of containing the 0.2 percent chance flood, a 40 percent chance of containing the 0.4 percent chance flood, a 85 percent chance of containing the 1 percent chance flood, a 98.5 percent chance of containing the 2 percent chance flood, and a 99 percent chance of containing the 4 percent chance flood.

In the case of the American River, variability in reservoir operation would also be added. Ford (personal communication, September 19, 1994); indicated that

> Reliability, as used in the American River study, describes the frequency with which a proposed plan performs as intended, given the occurrence of a specified event. It is computed via sampling also, with sampling of the frequency function limited to discharge of a specified exceedance probability. For example, the reliability of an alternative at the 1%-chance event is predicted by repeated sampling of the discharge and stage for a 1%-chance exceedance event (accounting for the error in predicting both the discharge and the stage).

Figure 4.5 illustrates the presentation of results for different project alternatives using the proposed reliability index. Numbers such as those in Table 4.4 appear in Figure 4.5 in graphical form so they can be better understood. An example is provided by Plate 12 in the 1994 Alternatives Report (Sacramento District, 1994).

In an early presentation of these concepts, Davis (1991) said

> One could (and likely would) prepare reliability tabulations of the likelihood of exceedance by various flood events to enable characterizing performance by assignment of a level-of-protection, with stated reliability.

> The approach suggested explicitly acknowledges that there is not a specific, unequivocal, performance level. Levels-of-protection would have to be couched in a reliability context such as "this levee project has a 95 percent chance of protecting against the 100-year exceedance interval flood, should it occur." Acceptable reliability criteria would have to be adopted.

Dotson et al. (1994) illustrated how the methodology can "quantify the reliability and performance" of a project, which can be expressed in statements such as "there is about a 95 percent chance of containing the 1-percent (100-year) flood, should it occur."

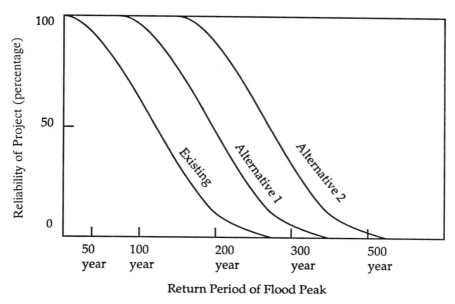

FIGURE 4.5 Illustration of the trade-off between the return period of a flood peak and the reliability of the reservoir-levee system with possible flood flows associated with that return period.

Application of Reliability Indices in the ARWI

While the committee does not disagree with the analysis in Table 4.4 or Plate 12 of the 1994 Alternatives Report, it cannot see clearly what the public or most engineers would do with such information. There are several concerns:

1. It is not at all clear how one should conceptualize the 1 percent chance event given that it is not converted into a single flow estimate. Instead it is used to generate a set of flows reflecting the hydrologic uncertainty in the computed discharge-frequency relationship. This makes it very hard to anchor the analysis mentally or to know for certain to what it is applied. In the definition of reliability for the American River study taken from Ford (personal communication, September 19, 1994, quoted above), what is the particular "specified event" to which the chance of failure in Figure 4.5 or in Plate 12 of the 1994 Alternative Report refers? Use of critical historic flood events with known flood flow peaks would help resolve this conceptual vagueness.

2. The Sacramento District needs to clarify its reasons for wanting to calculate this reliability index shown in Table 4.4. If the overall probability of levee

failure, which describes the residual risk of flooding, is already known for levees of different heights, what does this other reliability calculation add?

3. The analysis does not indicate how much of the reliability (or likelihood of failure) is due to hydrologic uncertainty, how much to stage-discharge uncertainty, and how much to variability in levee failure stage. It is not clear what this reliability calculation reflects.

The term reliability gave the committee the sense that it was a measure of how certain the Sacramento District is that the levee system would perform as intended. It was suggested that by using the new "reliability" index the Sacramento District is trying to tell the public that there is some uncertainty about how particular aspects of the project will perform. Ford (personal communication, September 19, 1994) wrote, "The term reliability, as used in the American River study, describes the likelihood that a proposed plan will perform as intended, given the occurrence of a specified event." Because the committee could find no explicit statement by the Sacramento District of what was intended, it had difficulty interpreting such statements. How should one define the intent of an existing system?

Because the numbers in Table 4.4, Plate 12, and Figure 4.5 also include the large uncertainty related to converting a "median" exceedance probability for a flood into the correct discharge, the committee found it very difficult to develop a useful interpretation of these numbers. It would be even more difficult for the public to interpret them. Proponents of careful risk communication warn of the pitfalls related to public misinterpretation of descriptions of risk (Plough and Krimsky, 1987; Slovic, 1987; NRC, 1989, 1994).

If each column in Table 4.4 corresponded to a discharge peak of a particular magnitude, corresponding to a historic flood or a selected design hydrograph, then one could interpret the calculated reliabilities as describing the consequences of stage-discharge estimation errors and levee performance uncertainty. For example, one could compute the reliability of the levee-reservoir system for a flood flow with a peak of 300,000 cfs or 500,000 cfs into Folsom Reservoir due to uncertainties in reservoir operation, stage-discharge relationships, and levee performance. One could also provide the estimated probabilities that these particular peak flows are exceeded. For levee systems, a similar calculation would result by having the columns in Table 4.4 represent particular discharges (and perhaps the estimated probability each would be exceeded).

4. Calculations of project reliability may involve some difficulties that are not apparent. In reservoir-levee projects the characteristics of a critical event may depend on the capacity of the reservoir(s) considered for different alternatives. For a levee-only system, it is the peak inflow that matters most. As one adds more storage, flood inflow volume becomes increasingly important. Thus one wants to select for the columns of a table such as Table 4.4, and for the graph

in Figure 4.5, events that for all alternatives are equally critical. It may be important that the public and the engineers who are reviewing project proposals understand how this is done. But this issue is completely hidden in a table such as Table 4.4, in which the character of the actual hydrologic event corresponding to each probability is obscure.

5. The explanation that is reproduced at the beginning of this section from EC 1105-2-205 also gives the sense that from Table 4.4 and Figure 4.5 one can determine the risk of flooding should a particular project be adopted. The residual risk of flooding is certainly a primary concern. Ford (personal communication, September 19, 1994) wrote,

> The importance of reliability is, to some extent, a function of the consequences of exceedance. If the consequences are great, then high reliability is necessary. For example, if overtopping a levee would inundate a high-density residential development to a depth of 25 feet without warning, high reliability is required.

This discussion of the importance of reliability ignores the risk associated with the target flow. The reliability of the system as calculated by USACE is only part of the residual risk. It is the overall risk of flooding that is key, not the reliability of the system for particular events. That overall residual risk of flooding, described by AFP or Avg[AFP], and the expected annual damages (EAD) are certainly the two most important system performance criteria.

A significant problem with the presentation of system reliability in Table 4.4 and Figure 4.5 is that reliability appears to address the residual risk of flooding, while it actually hides the true answer in a matrix of less meaningful numbers. To compute the actual risk of flooding (as described by Avg[AFP]), one would need to compute the average across all failure probabilities of the reliability of the system. There are insufficient numbers in the table to do this computation, and interested individuals should not have to do it themselves. Engineers and planners should perform these calculations and provide the results.

Table 4.4 and Figure 4.5 appear to reflect a desire to hold on to the old idea of "level of protection," expressed by the hydrologic return period T or exceedance probability for a design flood, while moving to a new risk analysis methodology that includes the idea of uncertainty and variability in other processes. Davis (1991) noted that traditionally projects were defined by the target "level of protection." The problem with the presentation of system reliability versus a target failure probability is that it fails to integrate those two sources of risk.

6. In the American River study, reliability is also used to demonstrate that the reliability of the levee network across the American-Sacramento River system is not impaired by a project. This is a legitimate concern and one that a risk analysis methodology should be able to address. The Sacramento District has demonstrated how its reliability index can be calculated at different points in the

system for different probability levels to demonstrate that "reliability" is not impaired.

The committee wondered if this is the most effective definition of reliability for that purpose. If reliability is used to ensure that for every flood level described by a cumulative probability p, throughout the river system the probability of flooding is not increased by a project, it is not necessary to include hydrologic discharge-frequency uncertainty. It is much simpler to specify a range of Folsom inflow hydrographs and evaluate the reliability for each. The "reliability of the system for a given inflow" is both simpler and more meaningful than the "reliability of the system for a given exceedance probability including our inability to determine the flow actually associated with that exceedance probability."

In this regard, requiring reliability to remain the same for every hydrograph is a more demanding requirement than requiring that it not decrease for every median exceedance probability after averaging over hydrologic uncertainty in the frequency curve. The first approach requires that reliability not decrease at every flood flow; the second requires that reliability not decrease for averages over flow ranges.

Keying on clearly specified flood hydrographs with their associated peak and volume seems to meet the requirement of ensuring that reliability not decrease more rigorously than the approach the Sacramento District has adopted. It would also be clearer and easier to understand. Moreover, it is also easier to compute and relate to levee and channel system performance because those uncertainties will not be swamped by the potentially much larger hydrologic uncertainty.

Overall, the committee applauds the USACE decision to adopt a risk-based planning methodology that better incorporates uncertainties in key variables. However, the committee does not believe that the definition for system reliability that was proposed in USACE guidelines and adopted by the Sacramento District is particularly effective at addressing the relevant issues. In many cases, it seemed unnecessary or misleading. Annual failure probability (AFP, Avg[AFP], or both) is likely to be the most straightforward and easily understood measure of residual flood risk. It could be supplemented by the vulnerability criteria discussed in the risk communication section of Chapter 6.

THE 1994 ALTERNATIVES REPORT

The committee reviewed the 1994 Alternatives Report (USACE, Sacramento District, 1994a) and found the document to be particularly confusing. The report provided a summary of its evaluation of different projects consisting of alternative modifications of the system. Unfortunately, essential details of the analysis were omitted, so the committee could not determine what was actually done from reading the report. In particular, the committee could not determine the extent to which some criticisms of the 1991 analyses had been addressed.

The report also failed to associate with the estimated net benefits any measure of overall uncertainty due to economic uncertainty, or hydrologic and hydraulic uncertainties, as recommended by EC 1105-2-205. These uncertainties could be important given the modest benefit-cost ratios calculated for the alternatives considered.

A very serious concern is how the report addressed issues of risk terminology and its reporting of flood risk. USACE now has two significantly different ways to calculate flood risk. On pages 8 and 9 of the 1994 Alternatives Report, they were both called "level of protection." No distinction was made between estimates of flood risk calculated with the traditional level of protection methodology and those calculated with the new risk and uncertainty methodology. Throughout the report a host of different terms and phrases were used interchangeably to describe these ideas. A layperson would have great difficulty sorting out the following jumble of terms: T-year level of protection, exceedance interval (p. 8), return period (p. 8), recurrence frequency (p. 9), control for T years (pp. 18 to 23), T-year flood (p. 9), T-year flood protection (p. 6), T-year protection (pp. 27, 29), T-year return frequency (p. 34), expected exceedance (pp. 37, 39), expected level of protection (p. 57), annual recurrence (Plate 5), and flood event return period (Plate 12). The report should use a few terms whose definitions are both clear and consistent with commonly accepted interpretations.

The most common terms in the report are T-year level of protection, T-year protection, control for T years, and T-year flood protection. The use of the term "level of protection" to describe flood risk is inconsistent with the new USACE guidelines for risk and uncertainty analyses (EC 1105-2-205) and confuses the traditional and the new approaches to calculating flood risk. This terminology supports the erroneous idea that one and only one T-year flood occurs every T years. Actual statements in the report reinforce the error. On page 9, flood risk was described as a flood once in 78 years or 103 years, while the executive summary indicated that "levees could fail about once in every 78 years" and "the level of protection (or likelihood that levees would not fail) would be increased to about once in 100 years." These are exactly the analogies that should be avoided.

With the new risk and uncertainty methodology, estimates of flood risk are no longer tied to a single T-year design flood, but can depend on different combinations of flood flows, operating decisions, and levee performance. Instead of stating that a project has a 200-year "level of protection" or protection for the 200-year flood, the Sacramento District should instead indicate that the annual risk of flooding is 0.5 percent per year, or the annual risk of flooding is 1 in 200 (see Stedinger et al., 1993, p. 18.3). It is also informative to convert such annual risks into the risk of flooding over 25 to 50 year periods, reflecting the likely length of a mortgage or the anticipated economic life of structures and dwellings.

When producing the 1994 Alternatives Report, the Sacramento District was under a great deal of pressure to revise its analysis of flood control alternatives to provide protection for people and property along the lower American River. Its difficulties were increased by the need to use the new risk and uncertainty methodology being developed within the Sacramento District for the planning of flood protection projects. Inadequacies in the 1994 Alternatives Report reflect those pressures and constraints. The committee hopes that subsequent documents will more clearly describe how the analyses were conducted and will more clearly explain the basis for the risk and uncertainty analyses.

THE PROMISE OF ECOLOGICAL RISK ASSESSMENT

USACE has made a commendable effort to apply recently developed risk and uncertainty analysis to the engineering problems faced in minimizing the damage from floods. The question then arises: Should not the relevant ecological risk and uncertainty that may be the consequence of each of the proposed actions also be subjected to risk analysis? Applying ecological risk assessment to the major areas of uncertainty would be a daunting task. The following discussion highlights some of the advantages and disadvantages of such an approach.

Development of the Paradigm

Formal assessment of risk in ecological science and management is a relatively new development. Until very recently, EPA had not developed any guidelines for risk assessment (Suter, 1993). Thus far, the principal application of risk assessment to ecological problems has been in the context of considering impacts of hazardous chemicals in the environment, evaluating the risk of extinction of rare or endangered species, or providing management advice for commercial fisheries. Conceptually, there seems to be no reason that the process could not be applied to assessment of potentially adverse effects of water projects such as the ones considered here in the American River. However, the extension to such an analysis is controversial (Lackey, 1994) and probably will not be generally accepted in the scientific community at this time.

Ecological risk assessment has evolved slowly over the past two decades, but has received impetus from the National Research Council (NRC) paradigm for human health risk assessment: *Risk Assessment in the Federal Government: Managing the Process* (NRC, 1983). EPA has recently released a *Framework for Ecological Risk Assessment* (EPA, 1992), along with a series of case studies (EPA, 1993, 1994). These publications do not present final policy and procedures but are designed to stimulate discussion and development of a process that will be in flux for some time. EPA is developing formal guidelines for conducting ecological risk assessments, which are expected to be released in late 1995 or early 1996.

NRC has been in the forefront of such development, with reports on risk communication (NRC, 1989) and on issues in risk assessment, including a significant discussion of ecological risk (NRC, 1993). Currently an NRC committee is conducting workshops designed to build consensus on the philosophy and methods for ecological risk analysis. In addition, many academic and industrial scientists are developing and evaluating the process (Suter, 1993).

The debate about the extension of risk analysis to ecological problems has focused on several contentious points:

- The process is based on a human health paradigm; extension to ecological effects, particularly at the ecosystem level, is highly problematic. There is insufficient understanding of ecosystem processes to predict outcomes with any certainty.

- Risk assessment has the potential to produce a sort of an ecological triage, whereby particular processes and species thought to be important might receive attention at the expense of some potentially serious problems.

- Risk analysis may lead to a consideration of alternatives that is too narrow, particularly if the focus is on the risk of a particular action versus that of no action. The analysis must consider the full range of alternatives, and benefits as well as risks of all the alternatives.

- The process can be tilted in favor of a particular action, given that uncertainty is great and the desired level of risk defined; the analysis may simply proceed until the desired endpoint is reached.

In spite of these serious concerns, ecological risk analysis has had some success, leading to models that may provide templates for further development. A recent NRC report (NRC, 1993), in a section titled "A Paradigm for Ecological Risk Assessment," recognized significant problems in extension of the health risk approach from NRC (1983). Nonetheless, that committee concluded that integrating ecological risk into the original framework is possible and that such an approach is preferable to developing a completely new framework. Key scientific issues limiting the application of ecological risk assessment include the following:

- Extrapolation across scales of space, time, and ecological organization. Estimating ecosystem-level response on the basis of laboratory or small-plot experiments is a particular concern.

- Quantification of uncertainty, including measurement uncertainty, natural variability in ecological systems, and inadequacy of models.

- Validation of predictive tools. Substantial improvements are needed in the models fundamental to effective risk assessment.

- Valuation of outcomes. Analysis of both costs and benefits is essential, but generally accepted principles for valuation of ecosystems are not available.

Ecological Risk Assessment and the American River

This committee believes that the Sacramento District has done a reasonably effective job of framing alternatives in the American River planning activities, particularly in the 1994 Alternatives Report. The recent organization of the Lower American River Task Force under the sponsorship of SAFCA has substantially improved communication among the various stakeholders in the basin. Hence there is the potential for appropriate use of ecological risk analysis. Nonetheless, there is little likelihood that such an analysis would be accepted by the scientific and lay community at this stage in the development of flood control proposals for the American River.

One of the most contentious environmental issues faced by the committee is the assessment of the potential effects in the canyons of the North and Middle Forks above a proposed detention dam at Auburn. Great uncertainty surrounds estimates of the probability of mass soil failure and mortality of vegetation following inundation. A case study example is available of risk analysis applied to a similar situation, modeling future losses of bottomland forest wetlands in Louisiana in the face of increased flooding (EPA, 1993). However, this analysis was based on a substantial body of research in that region and on the application of a simulation model adapted for the specific area. No such base of knowledge is available for the American River canyon. Scientific understanding that would allow accurate modeling of the processes involved in hillslope failure and mortality of vegetation is simply not available at this time, and most likely will not be available for years. Significant opportunities were missed when research failed to take advantage of the presence of the cofferdam upstream of the Auburn dam site, though the detention dam concept was not developed until after the dam breached in 1986.

One field of resource management has a relatively long history of recognizing uncertainty and may have lessons to provide as ecological risk assessment develops. Managers of marine and anadromous fisheries have long faced uncertainty. Stimulus for the development of more robust approaches to prediction in the face of incomplete knowledge has often come from the collapse of large fisheries (Ludwig et al., 1993). The model of adaptive management advocated by Holling (1978) and Walters (1986) recognizes that uncertainty is a pervasive element of most resource management scenarios. The committee strongly recommends that the water resource issues in the American River be managed in this adaptive context. Some important characteristics of this approach include the following:

- recognizing and communicating uncertainty,
- treating management as an experiment, and
- providing sufficient monitoring to allow managers to learn from the experience gained from observing system behavior.

The current direction in models being developed for management advice emphasizes Bayesian analysis (Walters, 1986; Hilborn and Walters, 1992) and statistical decision theory (Frederick and Peterman, 1995). Another trend in more traditional statistical analysis of natural resource issues has been a focus on statistical power analysis, particularly in the analysis of downward trends in resource abundance (Peterman, 1990). The approach has promoted more explicit consideration of where the burden of proof properly lies. Incorporation of these concepts into ecological risk analysis should improve future decisions in a wide array of resource conflicts.

CONCLUSION

From its review of the material provided describing the new USACE risk and uncertainty analysis guidelines, and the 1994 Alternatives Report, the committee reached the following conclusions.

• **Improvements in Planning Methodology.** The USACE risk and uncertainty methodology is an innovative and timely development. The explicit recognition of modeling uncertainty should result in a better understanding of the accuracy of flood risk and damage reduction estimates. The committee applauds the USACE efforts to develop a better flood protection planning methodology incorporating both risk and uncertainty in hydrologic, hydraulic, and economic parameters and processes. However, USACE and the Sacramento District need to more carefully develop and articulate the structure of their risk and uncertainty methodology, employing an effective vocabulary for distinguishing among risk, variability, uncertainty, and system reliability for use with technical and public audiences. USACE leadership is encouraged to convene an intra-agency workshop, including outside experts, to review the risk and uncertainty procedures, with special attention to the committee's concerns, and to recommend specific changes to the guidelines as necessary.

• **Impact of Uncertainty on Performance Criteria.** The proposed USACE risk and uncertainty methodology, which directly includes hydrologic uncertainties (and potentially other sources of uncertainty) in the calculation of average flood risk and the average annual flood damages that might be averted by a project, inflates those estimates. This upward bias is a concern if the methodology is adopted nationwide because it could distort the economic evaluation of projects. The committee did not have the resources to determine the actual distortion for the American River study.

• **Descriptions of Project Performance.** To avoid the problem of bias described in the recommendation above, and to simplify the analysis so that it can be more easily understood and is less dependent on hidden assumptions, the committee recommends that the primary descriptions of the expected annual flood damages and of the probability of flooding be based on best estimates of the parameters of models defining the deterministic and significant random processes contributing to flood risk and flood damage.

 • **Descriptions of Project Performance Uncertainty.** Best estimates of expected annual flood damages and the risk of flooding should be supplemented by descriptions of their uncertainty due to hydrologic, hydraulic, and economic uncertainties. Uncertainty can be described by a standard error or the distribution of the likely values of the quantity of concern. The impact of uncertainty can be illustrated by computing the probability that the national economic development objective is negative, or various quantiles of its distribution. The approach should be consistent with the requirement in USACE guidelines for risk and uncertainty analyses (EC 1105-2-205) that the estimate of NED benefits be reported both as a single expected value and on a probabilistic basis (value of the benefit and its associated probability) for each planning alternative. It is the committee's understanding that the American River study will not address economic uncertainties.

 • **Measures of System Performance Reliability.** Estimates of expected annual flood damages and economic benefits associated with different projects, and the probability of flooding at different locations, are likely to be the primary criteria describing flood risk and economic impacts. It will often be useful to calculate other indices of system performance and the reliability of different components of the river channel and levee system. The committee questions in general the value of the system reliability index proposed by USACE documents and employed by the Sacramento District in the American River study. It seems to be an awkward combination of traditional and new concepts.

 In the case of the American River study, a reliability index did have an important role in demonstrating that different projects do not increase the risk of flooding in any reach of the American-Sacramento River system. Still, it is not clear that the adopted definition is the most effective or easily understood. However, the Sacramento District's use of reliability does not affect the validity or accuracy of the study results and the calculations upon which they are based.

 • **Risk Analysis in USACE Alternatives Report.** The committee reviewed the risk and uncertainty analysis in the 1994 Alternatives Report. The report failed to associate with the estimated net benefits any measure of overall uncertainty due to economic, hydrologic, and hydraulic uncertainties. The committee found the explanation and presentation of the results particularly confusing. No distinction was made between estimates of flood risk calculated with the traditional level-of-protection methodology and those calculated with the new risk and uncertainty methodology. Both were called "level of protection" and described by a variety of terms, which further contributed to the confusion.

 The most common terms in the report are control for T years, T-year level of protection, and T-year flood protection. The use of the term level of protection to describe flood risk is inconsistent with the new USACE guidelines for risk and uncertainty analyses (EC 1105-2-205) and confuses the traditional and the new approaches to calculating flood risk.

This terminology and phases appearing in the report fosters the erroneous idea that one and only one T-year flood occurs every T years. Moreover, with the new USACE risk and uncertainty methodology that was employed, failure is no longer related to a single T-year design flood being exceeded, but can depend on different combinations of flood flows, operating decisions, and levee performance. Instead of stating that a project has a 200-year level of protection, or protection for the 200-year flood, studies should indicate that the risk of flooding is 0.5 percent per year, or equivalently that the chance of flooding is 1 in 200 each year.

• **The Promise of Ecological Risk Assessment.** At this time the committee does not believe that the process of ecological risk analysis is sufficiently evolved, nor that there is sufficient knowledge of the ecological system, for this new tool to be applied usefully to problems of flood control in the American River basin. However, ecological risk assessment does provide a new approach that emphasizes the importance of uncertainty in the analysis of the consequences of various alternatives. The process will help select questions for investigation and will be increasingly important in broadening the scope of future planning. USACE should follow this rapidly evolving approach and adopt it as soon as it shows promise of improving the decisionmaking process.

5

Flood Risk Management Behind Levees

THE FLOOD PROTECTION/DEVELOPMENT SPIRAL

Federally subsidized flood control projects, including upstream storage and local levee projects, were the prevalent form of national response to flood hazard between the mid-1930s and the late 1960s. Nevertheless, flood losses—one measure of flood hazard—continued to rise throughout this period despite the spending of billions of dollars to store and deflect floodwaters. In 1960, Gilbert F. White warned: "Probably the most important reason for the rising trend in flood losses [as measured in constant dollars] is to be found in the continuing encroachment of human occupance upon floodplains" (White, 1986). The 1966 *Report of the Task Force on Federal Flood Control Policy*, also authored by White, further charged that ". . . some flood plain encroachment is undertaken in ignorance of the hazard, that some occurs in anticipation of further federal protection, and that some takes place because it is profitable for private owners even though it imposes heavy burdens on society" (U.S. Congress, 1966). The trend of rising flood losses continues today.

Average annual losses have continued to rise in protected floodplains as well as unprotected ones. This occurs in part because construction of flood control projects can engender an illusion of total protection—a false sense of security—that leads to new growth within the area (White, 1975). If unrestricted development is allowed in the floodplain behind or downstream from a flood protection project, the overall value at risk becomes much greater than would have been likely in the absence of such protection. Although damage from moderate floods may be successfully averted, any structural project may fail owing to a design

many property owners were able to purchase flood insurance and later to receive claims payments. Other property owners did not have flood insurance or did not meet the 5-day waiting period for coverage. The Committee identified at least 67 flood insurance claims payments behind the Monarch Levee that totaled $13.2 million. This represents nearly 5 percent of the total flood insurance payments for the 9-state region. The flooding of this industrial area had severe impacts to the area not only from insured and uninsured damages but also from the temporary or permanent loss of jobs.

The federal task force recommended that FEMA not exempt areas behind levees from floodplain management and flood insurance requirements, unless the levees are designed to protect against the "Standard Project Flood," which, as established by USACE, normally exceeds the 100-year flood level. The task force report stated:

Action 9.6 Require actuarial-based flood insurance behind all levees that provide protection less than the standard project flood.

FEMA should designate as (flood restoration) zones those areas behind levees designed to meet current minimum NFIP criteria but that do not provide protection from the Standard Project Flood (SPF). The mandatory flood insurance purchase requirement would apply within this AL zone, and new buildings would pay flood insurance premiums based on actuarial rates. FEMA could establish floodplain management requirements for these areas, although elevation or floodproofing to or above the 100-year flood elevation should not be mandatory. This recommendation is similar to one in the 1982 National Academy of Science's National Research Council report, A Levee Policy for the National Flood Insurance Program. (p. 135 of final report)

The Task Force also recommended that the NFIP 5-day waiting period be increased to at least 15 days to avoid the surge of last-minute enrollments for flood insurance coverage experienced in the Midwest Floods. (FEMA subsequently raised the waiting period to 30 days.)

fluence of the American River and the Sacramento River at the city of Sacramento. In its natural state, the basin lies entirely within the 100-year floodplains of those rivers and associated local drainage systems. Today, the basin is physically bounded by a 41-mile ring of levees bordering the American River to the south, the Sacramento River to the west, the Natomas Cross Canal to the north, and the Natomas East Main Drainage Canal on the east.

Reclamation of the basin for agriculture began with various local levee schemes during the late nineteenth century and more formally dates back to the Federal Flood Control Act of 1917. Currently, most of the basin is devoted to agriculture, including extensive areas of irrigated rice cultivation and other grain crops on drier soils. Existing levees and agricultural drainage within the basin are operated by Reclamation District 1000. That district is a member of the Sacra-

TABLE 5.1 Population Growth of Sacramento Metropolitan Statistical Area and the City of Sacramento, 1970 to 1990

	1970	1980	1990	1980 to 1990 Change
Sacramento MSA	848,000	1,100,000	1,481,000	34.7%
Sacramento City	257,000	276,000	369,000	34.0%

SOURCE: Bureau of the Census, 1991.

mento Area Flood Control Agency (SAFCA), formed in 1986 to serve as a regional voice and catalyst for upgrading of flood protection in the Sacramento region.

The basin lies partly within the Sacramento Metropolitan Statistical Area (MSA) and also partly within the incorporated area of the City of Sacramento. Both the MSA and the city have experienced rapid development and population growth since 1970 (Table 5.1).

Local political jurisdiction over Natomas is divided among the city of Sacramento, Sacramento County, and Sutter County. Existing urban development in the basin occupies about 7,200 acres in the southern part of the basin between the American River and Interstate 80, which extends diagonally across the lower part of Natomas Basin. This developed area, known locally as South Natomas, lies within the city limits of Sacramento and directly across the American River from the city's business and governmental center. It includes about 13,700 structures and has a resident population of about 31,000. Interstates 5 and 80, two of the nation's primary highways, intersect at the northern edge of South Natomas.

An area of about 47,600 acres (75 square miles—larger than the District of Columbia), which makes up 86 percent of Natomas Basin, remains undeveloped at this time. Ultimately, full build-out of the basin could add 170,000 residents, with total development value exceeding $15 billion (Sacramento Bee, 1993a). The portion of the basin of immediate interest to Sacramento is a 7,000-acre area of agricultural land known as North Natomas directly to the north of Interstate 80. The city's general plan designates North Natomas as a major growth area for new housing and commercial development. It is projected to account for about 35 percent of new housing and 30 percent of new jobs in the city when fully built-out (City of Sacramento, 1993b).

Pressure to develop the Natomas area began in the early 1980s with proposals for industrial and commercial projects. These were opposed by the small group of existing residents in South Natomas, who sought more balanced, better planned new growth. In particular, they desired to upgrade their own area from a moderate income neighborhood to a more upscale locale.

Location provides the primary impetus for developing North Natomas. The area adjoins two interstate highways and is a few minutes of driving time from both downtown Sacramento and the airport. It would adjoin and expand on the present development in South Natomas. Under the North Natomas Community Plan, adopted by the Sacramento City Council on May 3, 1994 (amending earlier plans from 1993 and 1986), development is planned to take the form of 14 neighborhood areas, served by light rail, schools, parks, and retail centers. The plan also calls for a town center to be anchored by a regional park and a sports complex (Sacramento Bee, 1994b). The plan has been developed over several years by the North Natomas Working Group, which includes city and county planning officials; the Environmental Council of Sacramento; the Natomas Community Association; and the North Natomas Landowners Association, representing the 10 largest ownerships in the area.

One objective of the North Natomas Community Plan is to promote the fulfillment of federal and state air quality goals through the provision of light rail and the creation of new housing and employment opportunities convenient to each other and the central city, potentially reducing the need for longer distance commuting by car. Also, like most central cities, Sacramento seeks to stem middle-class flight and the loss of taxes and jobs to outlying areas beyond its jurisdiction.

As of mid-1994 the only urban facilities in Natomas Basin north of Interstate 80 consisted of the Sacramento International Airport in unincorporated Sacramento County and the Arco Arena, standing amid the still open fields of North Natomas. The sports arena was built in the mid-1980s by one of the area's major landowners to accommodate Sacramento's newly acquired professional basketball team, and to prime the pump of development in North Natomas. Since then, however, further development in North Natomas and throughout the Natomas Basin has stalled for a variety of reasons. Among these have been the national and state recession, unresolved questions of financing proposed infrastructure in North Natomas, and the need to develop an approved habitat conservation plan to protect the Swainson's Hawk and the Giant Garter Snake (state-listed threatened species found in the Natomas Basin). But overshadowing other impediments to development has been the unresolved issue of flood hazard.

Natomas Flood Hazard

Although lessened to some degree by its 41-mile ring of levees, the risk of flooding in the Natomas Basin remains significant. Much of the land surface of the basin lies below the levels of the American and Sacramento rivers at flood stage, and also below the elevation of the surrounding land. According to USACE: "flooding from levee failure would be similar in Natomas, downtown Sacramento, and to some extent North Sacramento *regardless of the frequency of the flood event* because: 1) the ground elevation adjacent to the levees in these locations is lower than the water surface in the river, and 2) the volume of water in the American River . . . and Sacramento River in the case of Natomas . . . would fill the flood plains to similar depths" (USACE, Sacramento District, 1991) (emphasis added).

In other words, the basin could fill like a bathtub to the level of the bordering rivers in the event of a flood exceeding the design capacity of the Natomas levees, or a lesser event that breaches weaknesses in the levees (as occurred widely in the Mississippi River floods of 1993). Critics point out that a flood of any frequency, if gaining access to the area within the levees, would inundate over 59 percent of the basin to a depth greater than 13 feet, 32 percent to a depth between 8 and 13 feet, and 9 percent to a depth of less than 8 feet (Estes, 1993). (See Figure 5.1.)

The level of protection provided by the existing Natomas levees is uncertain. The American River levees in conjunction with Folsom Dam upstream were thought to provide approximately 100-year protection until the flood of February 1986 compelled reconsideration of that assumption. Releases from Folsom Dam in that event reached a peak flow of 132,000 cubic feet per second (cfs), which exceeded the design target of 115,000 cfs of the downstream levees. The levees were not overtopped, but the flow encroached into the design freeboard and caused erosion in certain areas. A few more hours of rain would have resulted in major disaster. The 1986 flood was rated as a 70- year event (City of Sacramento, 1990). USACE subsequently downrated the available level of protection along the American River below Folsom Dam to 40 to 70 years. The Sacramento River levees bordering Natomas were estimated by USACE after the 1986 flood to protect only against a 40-year flood (City of Sacramento, 1990).

Repair and renovation of existing levees along both the American and the Sacramento rivers have been in progress since 1986. Congress authorized further upgrading of the levees and internal pumping and drainage facilities for Natomas in 1992 (in P.L. 102-396, which also created this committee), as recommended in the 1991 ARWI. This congressional authority, however, was subject to the condition "that such construction does not encourage the development of deep floodplains" (Section 9159(b)(1)). (The act did not define the term "deep floodplains.")

Even without a levee failure, internal drainage remains a serious problem for development in the Natomas Basin. Mechanical pumping systems are inadequate to remove surface drainage when river levels are higher than the land surface of the basin, so local shallow flooding may occur even during storm events that do not threaten the levee system.

NFIP Status of Natomas

The development of the Natomas Basin is further affected by its novel status under the National Flood Insurance Program (NFIP). Normally, where new construction is built in "special flood hazard areas" (100-year floodplains mapped

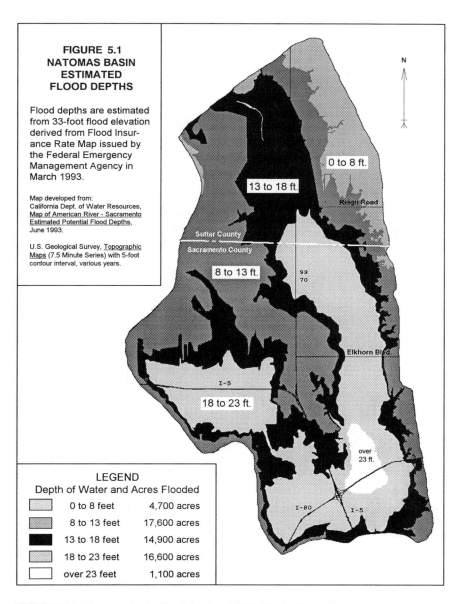

**FIGURE 5.1
NATOMAS BASIN
ESTIMATED
FLOOD DEPTHS**

Flood depths are estimated
from 33-foot flood elevation
derived from Flood Insur-
ance Rate Map issued by
the Federal Emergency
Management Agency in
March 1993.

Map developed from:
California Dept. of Water Resources,
Map of American River - Sacramento
Estimated Potential Flood Depths,
June 1993.

U.S. Geological Survey, Topographic
Maps (7.5 Minute Series) with 5-foot
contour interval, various years.

0 to 8 ft.

13 to 18 ft.

Reigo Road

Sutter County
Sacramento County

8 to 13 ft.

Elkhorn Blvd.

I-5

18 to 23 ft.

over
23 ft.

**LEGEND
Depth of Water and Acres Flooded**

	0 to 8 feet	4,700 acres
	8 to 13 feet	17,600 acres
	13 to 18 feet	14,900 acres
	18 to 23 feet	16,600 acres
	over 23 feet	1,100 acres

FIGURE 5.1 Natomas basin flood depths. SOURCE: G. Estes, 1993.

by NFIP), NFIP (1) requires residential structures to be elevated at least 1 foot above the estimated 100-year flood level and nonresidential structures to be floodproofed or elevated to that level; (2) requires owners to purchase flood insurance if they borrow money from a "federally related source" for purchase or construction of structures in such areas; and (3) charges "actuarial rates" for flood insurance coverage reflecting the actual risk of flood damage at the site in question. Outside of special flood hazard areas, NFIP does not require elevation or insurance purchase, but makes coverage available in eligible communities at nominal rates.

Areas protected by levees certified to provide "100-year" protection, are considered by NFIP as *not* within a special flood hazard area, although floods exceeding the design standards of the levees may still inflict catastrophic damage. Natomas was thus exempted from NFIP requirements until the downrating of its levees after the 1986 flood. In 1988, special legislation attached to the McKinney Homeless Assistance Act (P.L. 100-628) specifically exempted Natomas and certain other floodplains in the Sacramento area from the imposition of NFIP elevation requirements for a period of 4 years. The rationale for this special exemption for Sacramento—the first in the history of the National Flood Insurance Program—was the concern that a change in flood elevation requirements " . . . will cause severe disruption in the Sacramento region and could precipitate the breakup of the political, institutional, and economic relationships sustaining the high level, comprehensive flood protection efforts" (P.L. 100-628, Section (a)5). The act further stated that "the City and County of Sacramento have each provided assurances to the Congress that they will not designate any increases in urbanization beyond lands already so designated in their general plans . . ." during the 4 year period of exemption (Section (a)7). Further, they committed themselves to consider (P.L. 100-628, Section (a)7):

a) an evacuation-emergency response plan;

b) mechanisms by which to attempt to provide notice to all buyers of new structures;

c) retention of natural floodways; and

d) recommendations to all buyers of new structures to purchase flood insurance.

The Federal Insurance Administrator sent a letter to Congressmen Vic Fazio and Robert T. Matsui that viewed the exemption with dismay (H. Duryea, personal communication, January 3, 1989):

> Although the statute does not directly address the issue of insurance rates, the estoppel on the establishment of new base flood elevations effectively prohibits FEMA [Federal Emergency Management Agency] from charging actuarial (risk-based) rates within the areas specified. Since risk data is available from the Corps of Engineers' study, but Congress has prohibited the establishment or

alteration of base flood elevations based upon it, the assignment of insurance rates becomes somewhat arbitrary. In light of this situation, FEMA will act in accordance with your position that the statute prohibits changes to insurance rates within these areas, and continue to make flood insurance available at rates normally utilized outside areas of special flood hazard.

I am sure you recognize that maintaining the status quo with regard to insurance rates in these areas will create a significant subsidy for new construction, as actuarial rates are likely to be considerably higher than the current rates. This subsidy will be further increased because FEMA will be unable to enforce the program requirement that new construction be elevated to base flood levels. This subsidy could exist well beyond the maximum 4-year moratorium period specified in the statute, if the anticipated structural flood control solutions are not realized in the near future. In this situation, other flood insurance policy holders and taxpayers in California and the remainder of the country would be paying for the expected flood losses to new construction which will be built in these areas during the next four years.

In compliance with the statute, FEMA remapped Natomas as an "A-99 Zone" within which federal floodplain management requirements are minimal. Lenders must notify borrowers of a potential flood risk, and the latter must purchase flood insurance, but rates are those applicable outside special flood hazard areas (i.e., rates are very low).

In fulfillment of their commitment to Congress, the City and County of Sacramento adopted a "Land Use Planning Policy Within the 100-Year Flood Plain" and its accompanying environmental impact report on February 6, 1990 (City of Sacramento, 1990). The policy (1) allows approved development in areas of the 100-year floodplain outside of Natomas; (2) imposes a de facto temporary moratorium on residential development in the Natomas area during the period of Sacramento River levee instability by conditioning building permits on compliance with regulations applicable to development in a flood hazard zone; (3) conditions all nonresidential building permits on compliance with structural design and planning criteria aimed at minimizing the risks due to flooding; and (4) reaffirms the city's commitment to Congress not to designate any increases in urbanization in floodplain areas beyond lands already so designated in the city's general plan (City of Sacramento, 1993b).

Congress in 1992 amended the National Flood Insurance Program to create a new "flood restoration zone" for areas where levees have been downrated but are being expanded to restore 100-year flood protection. Within a flood restoration zone, modified NFIP requirements would apply to areas (1) that previously were accredited by FEMA as protected by levees to the 100 year level; (2) whose flood protection had been decertified by a federal agency, and (3) that are in the process of restoring such 100-year protection using federal funds (P.L. 102-550, Section 928, amending Section 1307 of the National Flood Insurance Act). Senator Alan Cranston, chair of the Senate Housing and Urban Affairs Subcommittee responsible for the bill, stated on the floor of the Senate: "The flood restoration zone is not intended to allow the development of undeveloped areas including the Natomas basin area in Sacramento" (Congressional Record, 1992).

Proposed regulations concerning the AR zone were published by FEMA on April 1, 1994, in the Federal Register (p. 15351-15361). Essentially, the amendment and proposed rules would limit mandatory elevation of new structures to 3 feet above grade in "developed areas" or where expected 100-year flood elevations would be 5 feet or less in "undeveloped" areas. Where expected elevations would exceed five feet in undeveloped areas, new construction would be elevated to the actual estimated level. No elevation requirements would be imposed on substantial improvements to existing structures in flood restoration zones.

However, the regulations also provide for "dual zone" designations (e.g., AR/A1-30, AR/AE, etc.) that reflect residual risk from flooding even when the levee system is restored to 100-year level of protection. In such dual zones, new construction and substantial improvements would have to be elevated as otherwise required for the underlying zone.

Mandatory purchase of insurance requirements under the Flood Disaster Protection Act of 1973 would apply within AR or dual zones. However, premiums for coverage would be charged at rates applicable to those for C zones (i.e., pre-FIRM subsidized rates). Eligibility for flood restoration zone status is limited to "communities where construction and restoration of a flood protection system is a Federally funded project and the existing flood protection system was constructed with Federal funds . . ." (Federal Register, April 1, 1994). The American River Project Levee bordering Natomas was constructed by USACE after the completion of Folsom Dam (U.S. Bureau of Reclamation, 1986), but current improvements are being conducted by SAFCA, a nonfederal agency. It is therefore unclear whether Natomas will qualify for flood restoration zone status, and if so whether portions of it will be designated as dual zones where elevation to applicable levels would continue to be required. Clearly, most of Natomas outside the existing development in South Natomas is currently agricultural and would be designated "undeveloped."

The city and county moratorium was partially lifted on October 12, 1993, by the City Council, which voted to allow master parcel plans to be approved for Natomas. That action would potentially allow developers to qualify for loans to finance infrastructure. Individual building permits, however, will not be issued until the federal government certifies that the area is protected against a 100-year flood (Sacramento Bee, 1993b).

The City Council, at the urging of Friends of the River, the Sierra Club, and the State Reclamation Board, on December 7, 1993, ordered a comprehensive floodplain management plan for the city to be prepared within 12 months (R. Stork, personal communication to R.H. Platt, May 19, 1994). The goals of this effort as defined by the Council (City of Sacramento, Department of Utilities, 1994 are to (1) . . . provide the areas designated for urban development within the

City of Sacramento with at least 100-year level of flood protection on a short term basis; (2) . . . provide the City with the highest level of flood protection, minimum 200 year, along the Sacramento and American Rivers on a long term basis; and (3) [utilize] all secondary flood protection measures needed to address the residual risk of uncontrolled flooding in Sacramento.

The long-term goal of "minimum 200-year" level of protection, while desirable, is unlikely to be achieved through reconstruction of local levees within the immediate control of SAFCA and its constituent local government agencies. This policy, as stated by the City Council, thus implies an expectation that 200+ year flood protection will be provided externally, through reoperation of Folsom Dam, through construction of an Auburn dam, or by other means. Indeed, recent recalculation of expected flood flows on the Sacramento River may even jeopardize the achievement of 100-year protection through levee improvements currently in progress (Sacramento Bee, 1994a).

The City Council has thus established a goal—200+ year protection—that is beyond the capability of the city to achieve within its own boundaries and through local actions. Yet it is poised to approve the development of North Natomas and other floodprone areas despite the unresolved, and perhaps unresolvable, issue of ongoing flood hazard. The city tends to blame external entities, principally Congress and federal agencies, for its inability to move ahead with development of Natomas free of any consideration of residual flood risk. In a report dated October 12, 1993, the city declared that the only obstacles to development of its floodplains were the lack of action by Congress and the uncertain future status of Natomas under the National Flood Insurance Program (City of Sacramento, 1993a).

> The Land Use Planning Policy EIR [approved 1990] assumed that by the time the Sacramento River levee stabilization project was completed in 1992, Congress would have authorized a long-term flood control project and all obstacles to achieving at least a 100 year level of protection for the Natomas area would be removed. Instead, Congress failed to authorize a long-term project and although SAFCA has proceeded with levee improvements . . . it is still uncertain when these improvements will be completed. Despite this uncertainty, Natomas' existing A-99 zone designation persists because no new zone has been designated to take its place. FEMA expects to provide a new designation by the end of 1994. If at that time, the Local Levee Project meets FEMA's adequate progress criteria, Natomas' existing A-99 zone designation could be formally extended. *This designation would allow building permit issuance without structural elevation.*
>
> However, if adequate progress criteria are not met, FEMA is likely to designate Natomas as a flood restoration zone. In that event, it is unclear whether and to what extent building permit issuance could proceed. Congress has provided in committee language attached to the AR zone legislation that this zone designation is specifically not intended to facilitate widespread development in Natomas.

CONCLUSION

Clearly, as a site for growth the Natomas Basin is well situated in terms of proximity to urban development, but it is poorly situated in terms of chronic flood risk. The recorded history of the Sacramento-American river system has been a long chronology of greater than expected floods, the latest of which in 1986 nearly overwhelmed the local flood protection facilities (Kelley, 1989). Improvements in the existing flood protection system including the reoperation of Folsom Dam and levee expansion are in progress or foreseeable. Other measures, such as construction of a dry flood storage dam at Auburn, are hypothetical and speculative at this writing. Environmental, fiscal, and political contingencies are likely to continue to delay that option and perhaps render it entirely moot. The future level of reliable flood protection therefore is difficult if not impossible to assess in light of both hydrologic and sociopolitical uncertainty.

Because of this continuing uncertainty, the committee makes the following recommendations: Whatever development proceeds within the 41-mile ring of levees surrounding the Natomas Basin should be subject to prudent floodplain management requirements under federal, state, and local authority. Unless the levees are certified to protect against a "standard project flood," the basin should be designated as subject to a residual flood risk on NFIP flood insurance rate maps. This may require the specification of a new "AL zone" classification as recommended by the Federal Interagency Floodplain Management Task Force. Within the area so designated, appropriate requirements concerning minimum elevation, building design, and mandatory insurance purchase should be administered. Flood warning and evacuation capabilities should be developed by responsible local authorities. In areas subject to possible deep flooding, the designation of in situ shelters in taller structures should be considered. The public should be informed of the flood risks that exist in the Natomas Basin despite the presence of the levee system. It is essential that federal agencies, and particularly the National Flood Insurance Program, not accede to local desires to develop the Natomas Basin under the illusion that the threat of flooding can be eliminated.

As stated by Doug Plasencia, chair of the Association of State Floodplain Managers, in testimony to the U.S. Senate Committee on Environment and Public Works, Subcommittee on Transportation and Infrastructure, on February 14, 1995:

> If we as a nation are going to bring escalating disaster costs under control, we need to embrace the concept of hazard mitigation. Our flood policies have not embraced hazard mitigation, and in part are to blame for escalating disaster damages. On an individual project basis, flood control projects have reduced flooding for design floods. But at the same time these policies have enticed additional development, increasing the damage potential for severe floods, or have silently promoted the transfer of flood damages from one property to another. Likewise, with a benevolent federal government, there has been little incentive at the local and state levels of government to minimize the creation of new flood hazards.

6
Flood Risk Management: Implications for the American River and the United States

Since the mid-1960s, escalating costs and environmental opposition have posed formidable barriers to the construction of new flood control projects (NRC, 1993). Alternatives to construction increasingly are being sought, including nonstructural measures to reduce exposure to flood damages and insurance to compensate for damages incurred. Thus Congress's reluctance to authorize the U.S. Army Corps of Engineers (USACE) recommended flood control dry dam on the American River (USACE, Sacramento District, 1991) and its request to continue studies of other alternatives to address the flooding problem (P.L. 102-396) are not unexpected.

In November 1994 the Sacramento District of the USACE published its Alternatives Report (USACE, Sacramento District, 1994a). Although providing only limited detail, the report suggested that the District has responded to past criticisms: it has reconsidered increased flood conveyance and environmental restoration opportunities on the lower American River, credited storage in upstream hydroelectric reservoirs for flood control under certain circumstances, and considered different assumptions about future Folsom flood operations. In addition, the Sacramento District has employed newly developed USACE risk analysis procedures to compare alternatives.

This chapter begins with a discussion of the committee's understanding of how the USACE planning and decisionmaking process, described in Chapter 1, was applied in the American River Watershed Investigation (ARWI). The discussion considers the source of the controversy over acceptable alternatives to address the American River flooding problem. Subsequent sections discuss the following aspects of the American River controversy in more detail:

- acceptable flood risk and the flood insurance program,
- water project cost sharing,
- communication of flood risk,
- improved approaches to flood risk management planning, and
- the water policy and management context.

The chapter concludes with recommendations for reforms to current planning and decisionmaking for the American River situation. Because the committee believes that the lessons of the American River can be transferred to other areas of the nation, parallel recommendations for reform of national policy on flood risk management are also offered.

THE AMERICAN RIVER FLOOD RISK MANAGEMENT CONTROVERSY: THE KEY ISSUES

The Flood Control Act of 1962 (P.L. 87-874) authorized USACE to study the American River basin in the interest of "flood control and allied purposes." However, the funding to execute this authority was not provided until after the 1986 flood. In providing funds for a one-year reconnaissance study, the committee language in the Fiscal Year 1988 Continuing Appropriations Act (P.L. 100-202) defined a broad scope for the studies, although the priority was on the imminent flood risk (USACE, Sacramento District, 1991, hereafter the 1991 ARWI):

> The conferees are aware that recent information presented by the Corps and the Bureau in a series of three fact-finding hearings in Sacramento reveals that the region may be under a greater threat from serious flooding than was previously believed. . . . Within this assessment, the Corps should include its analysis of the current and projected water supply demands in the American River basin.

Most USACE study authorities mandate that multiple purposes, including flood control, navigation, hydroelectric power, water supply, recreation and fish and wildlife habitat improvements, be addressed. The initial planning task is to focus limited study resources on the most pressing planning problems and opportunities, and the congressional directions for the American River clearly pointed to flood damage reduction through flood control as the priority. However, the public comment record also indicated a strong interest in ensuring reliability and reducing costs of water supply, in increasing hydroelectric power generating capacity, in promoting restoration of environmental resources that had been degraded by past water development, and in protecting and enhancing recreational opportunities (USACE, Sacramento District, 1991, Appendix T). Indeed, many who commented were critical of the Sacramento District's 1991 ARWI for its failure to consider any purpose other than flood control.

With its planning attention focused on flood control, the Sacramento District proceeded to examine the alternatives for reducing flood risk. Sound planning practice demands that the range of alternatives include combinations of engineering, regulatory, and other public policy measures to, in this case, provide an acceptable degree of flood damage reduction. For example, alternatives for meeting the flood damage reduction purpose might include different sizes of reservoir storage, different restrictions on floodplain settlement, and different levee heights, all in different combinations. Many critics of the 1991 ARWI felt that two sizes of dry dam, each designed without gates to foreclose the option of permanent storage, received favored attention. Ironically, some chastised the District for its focus on the dry dam because these critics wanted a more extensive consideration of a new full-pool dam that would address the flood damage reduction as well as provide for many other purposes.

It is worth noting that USACE was instructed by language accompanying the Fiscal Year 1988 Continuing Appropriations Act (P.L. 100-202) to pay special attention to flood control through a dry dam option (USACE, Sacramento District, 1991):

> The conferees . . . recognize that there may be additional flood protection afforded by a primarily peak-flow flood control facility (the so-called "dry dam") on the North Fork of the American River above Folsom. The conferees therefore direct the Corps of Engineers to include further assessments of the relationship between such a peak-flow flood control facility and the operation of Folsom Dam as they may pertain to incidental water, power and recreational benefits.

The dissatisfaction with the limited purposes and alternatives considered by the Sacramento District contributed, in part, to the failure of Congress to authorize the dry dam proposal. First, even though the financial and environmental impediments to implementing the full-pool alternative—akin to Bureau of Reclamation's originally proposed Auburn dam—were formidable, continued support of the full-pool alternative is found among water supply interests. The committee was told of "foothills" communities east of Sacramento (e.g., those within the El Dorado Irrigation District) who see a full-pool alternative as a means to ensure their access to water rights and to reduce their future water supply pumping costs. Indeed, some of the SAFCA Board of Directors may still favor a full-pool alternative for the water supply purpose. A dry dam that appears to foreclose water supply expansion will have limited support among these interests.

Second, the dry dam also was opposed by canyon protection interests concerned about the effects of occasional impoundment on soils, vegetation, and wildlife in the American River canyon (see Chapter 3). Furthermore, canyon protection interests worry that a dry dam will in time be converted to a permanent pool. In opposing the dry dam, canyon protection interests openly state this concern. By 1991, both those who opposed any dam in the canyon and those who

favored a full-pool multipurpose Auburn reservoir had expressed strong dissatisfaction with the dry dam alternative.

The dispute over the Auburn dam proposals has stalled the entire lower
American River study process. Recent planning efforts have sought to match
environmental restoration concerns with improved levee stability and conveyance in the Lower American River Parkway corridor. Agreement appears to be
near under the auspices of SAFCA's Lower American River Task Force (SAFCA,
1994a). However, there seems to be little progress on resolving disputes over
storage in the canyon. In the 1994 Alternatives Report, which has reanalyzed the
data, the Sacramento District continued to report that substantially reduced flood
risk and increased net benefits could be achieved by construction of storage at the
Auburn site. Therefore, a question remains: Can an alternative based on existing
storage and on the lower American River levees provide an acceptable level of
risk reduction for the city? The ongoing reevaluation of alternatives is expected
to help resolve this question.

USACE evaluation of alternatives is governed by the Water Resources
Council's *Principles and Guidelines* (P&G) (Water Resources Council, 1983).
The P&G requires the federal agency to recommend the alternative that makes
the greatest contribution to national economic development (NED). This alternative maximizes net benefits over costs. The NED plan also must comply with
national environmental statutes, applicable executive orders, and other federal
planning requirements. The new USACE risk analysis procedures do not change
the requirement to develop an NED plan. As is noted in Chapter 4, the procedures are intended to provide a more realistic and complete description of the risk
and uncertainty associated with reservoir and levee system performance, and the
associated flood damage reduction benefits.

The estimates of the value of the benefits needed to ascertain the NED plan
are obtained in various ways. For flood damage reduction, the avoided future
repair costs to property no longer exposed to flooding is the most widely employed benefit measure. Another measure of flood control benefits is the change
in property prices with versus without a flood control alternative. Other benefits
might be estimated for increased recreational opportunities or enhanced water
supply. The alternatives in the 1991 ARWI were compared primarily in terms of
flood control benefits.

Included in the flood control justification for the dry dam alternative were
benefits to be accrued for the still-to-be-developed Natomas area. Critics of the
1991 ARWI noted that this placed the federal government in the position of
building a project that would facilitate the development of a flood-prone area, a
position they deemed unacceptable. However, future growth in Natomas made
up only a small percentage of the total benefits, so removing these benefits from
the NED justification did not make any dry dam alternative economically unjus-
tified. Nonetheless, it is likely that the political support for a dry dam, given the

desire to retain the development near the city, may have been influenced by the prospective development at Natomas (see Chapter 5).

The costs in an NED analysis include public and private sector investment and future operation and maintenance spending, as well as monetary values for foregone hydroelectric power generation or water supply. The P&G requirement to comply with applicable environmental laws, which has been interpreted by USACE to include full mitigation of any adverse environmental effects of an alternative, adds to project cost.

The uncertain frequency, depth, and duration of inundation behind a dry dam called into question the adequacy of the mitigation offered for damages in the canyon. If the mitigation was not adequate, then the costs of mitigation were understated. This concern, which is discussed in Chapter 3, was the central environmental challenge to the dry dam and was the major mitigation feature of any of the alternatives. However, criticisms of the many alternatives also have been related to concerns about negative environmental effects or foregone opportunities for environmental restoration along the lower American River.

Preliminary cost estimates for the array of all alternatives were presented in the 1994 Alternatives Report. Average annual costs (at an 8 percent discount rate) range from $22 million for the minimum environmental impact plan having first costs (up-front construction costs) of $258 million, to $68 million for the dry dam designated as the 1991 NED alternative. In the 1991 ARWI, the first cost estimate for the dry dam was $698 million and in the 1994 Alternatives Report it was $661 million. The foregone benefits of water supply, power, and recreation from Folsom reoperation are presented as $4 million per year. It appears from the available information that these costs were estimated as the costs to construct new storage ($300 per acre-foot) to replace these benefits at another site (USACE, Sacramento District, 1991). If this was the estimation method, then this cost estimate is an upper bound because replacement water supply, for example, might be acquired from water markets at lower costs (Science Applications International, 1991).

The project costs are shared between levels of government according to formulas fixed by federal law. The 1994 Alternatives Report did not include any cost sharing information; however, the 1991 ARWI indicated that the 1992 cost of the selected plan would have required a $240.5 million local contribution, of which $208 million would have been in contributions of lands, easements, rights of way, and relocation. The most significant share of the nonfederal cost was $107 million to relocate Highway 49. This cost would be borne by the state, leaving $101 million as the local cost for the project (USACE, Sacramento District, 1991).

For a flood control project, it is important to estimate the likelihood that an area will be inundated. The "level of protection" has been used to indicate a likelihood of flooding in any year. For ease of exposition, the reciprocal of the annual likelihood—that is, the average number of years between occurrences, or recurrence interval—is commonly used. For example, a flood discharge with a 1-percent chance of being exceeded in any year—1/100—is referred to as the 100-year flood. (A discussion of the meaning and limitations of the term level of protection is found in Chapter 4 and later in this chapter.)

The level of protection that is to be recommended by USACE policy tends to coincide with whatever can be achieved by the NED alternative. Thus the NED plan can result in different amounts of residual flood risk for different studies, depending on the site-specific costs and benefits of flood damage reduction. The NED plan in the 1991 ARWI offered a 400-year level of protection. However, SAFCA, for reasons of cost and other considerations, identified a 200-year level of protection *from a dry dam alternative* as the locally preferred plan. The SAFCA choice was accommodated by USACE budget policy that permits selecting an alternative other than the NED plan if, in the words of the P&G, ". . . there are overriding reasons for recommending another plan, based on other federal, state, local, and international concerns." The SAFCA preferred plan was recommended.

Those who were concerned about threats to the canyon asserted that an acceptable level of protection could be achieved with levee elevation, reoperation of Folsom, and other management measures, without any new storage facilities. Critics argued that the only reason the dry dam was supported was to ensure protection for new development in "deep floodplains" at Natomas. The critics cast the choice as between "saving" the canyon and serving speculative land development. However, some supporters of the dry dam felt that any alternative with no storage was less reliable because of uncertainties in the hydrologic modeling, the likely effectiveness of reoperation of Folsom, and the structural condition of the levees. In fact, the new USACE risk analysis procedures were developed to directly address and analyze such uncertainty (see Chapter 4).

The Sacramento District favored the 400-year protection dry dam alternative, SAFCA sought 200-year protection through a dry dam, and others argued that an acceptable level of protection was possible without a dry dam. Yet a fourth perspective was that a level of protection only against the 100-year flood was required by federal policy. EPA, for instance, cited FEMA flood insurance purchase requirements as evidence that this was a national standard (Wieman, 1992). USACE policy should, according to EPA, be to develop alternatives that equally achieved the minimum "protection" against the 100-year flood. Then the "least environmentally damaging" alternative to meet that standard should be chosen. This decision logic, rooted in rules of Section 404 (b)1 of the Clean Water Act, was rejected by USACE as binding on their formal planning.

The 1991 ARWI study generated hundreds of letters and much interagency comment. After considering this public agency input, the Sacramento District finally supported the SAFCA preferred alternative, but that recommendation was rejected by Congress at the urging of environmental and water supply interests. Clearly, an "open comment process" did not satisfy those with the ability to block implementation of a dry dam. Although local cost-sharing requirements certainly required the district give serious consideration to the local project sponsor, in this case SAFCA did not represent the myriad regional concerns regarding development priorities and environmental issues. The planning process in 1991 was preoccupied with reconciling differences between SAFCA and USACE while other interests were left to exercise political strategies outside the planning process, ultimately with much success.

The proposed dry dam project only partially fulfilled the desires of the local sponsor and did not meet the NED standard of the P&G. Yet even this compromised project was opposed by an unusual combination of interests—those who sought to stop any project in the canyon and those who wanted a full-pool reservoir. Although the two had diametrically opposed positions, neither stood to satisfy its preferences if the 1991 ARWI preferred alternative were implemented. Stopping the SAFCA preferred plan in order to fight for different alternatives in the future must have appeared to each group to be in its own interest.

As of March 1995, it did not appear that these positions had changed significantly. Canyon protection interests have continued to assert that no solution that includes a dry dam could be implemented, but the support for a multipurpose dam had been restated by other interests (Sacramento News and Report, 1995). Meanwhile, SAFCA has continued to advance its preference for a 200-year level of protection, whether or not involving a dry dam.

The current decisionmaking situation in the American River basin can be described as a diffusion of separate interests having access to numerous political and legal veto points. This situation is not unusual. USACE has found in recent years that its recommendations are frequently challenged, often with success. The committee thus decided that it was important to comment on selected aspects of the planning and decision-making process, as well as to recommend reforms for federal policy on flood risk, with the expectation that such observations will contribute to reaching a decision in the American River basin.

THE CHOICE TO BE MADE:
ACCEPTABLE REMAINING FLOOD RISK

The committee has not identified any national standard for acceptable levels of flood risk reduction, but it understands the intent of current national policy as follows: investment (private or public) in hazardous areas should, to the extent practicable, internalize the costs of choosing such a location by (1) contributing to the cost of floodwater control works, (2) accepting reasonable restrictions on development in flood-prone areas (foregone development value), and (3) paying adequate insurance premiums against the flood risk remaining after structural and nonstructural measures have been implemented.

The committee recognizes that settlement of floodprone areas may bring advantages in terms of access to urban services, opportunities, and amenities (as in the case of the Natomas area—see Chapter 5). However, the committee also believes that such advantages must be balanced against the threat of water inundating a settled area and the damages or loss of life related to the depth of flooding, the velocity of the flow, and the rapidity and duration of the inundation. Such balancing is the essence of flood risk management. Flood risk management decisions are made by individuals and communities when they choose to locate economic activity in flood-prone areas and when they choose to implement particular measures that will reduce the frequency of inundation in an area or the susceptibility to damages from any inundation that does occur.

Flood risk in any year can be described by the expected damages from each of the possible inundation events, weighted by the likelihood each event will occur. Flood risk management decisions weigh the avoided damages (benefits) of flood risk reduction against the cost of alternatives for reducing flood risk. Costs of flood risk management are the budget outlays for a flood control project, plus any unmitigated environmental damages from the project. Costs also include the foregone value of activities either removed from, or not located in, a flood-prone area. However, in few instances will the benefits of removing all flood risk justify the costs of an alternative to achieve those benefits. For the American River, for example, flood risk management requires deciding the "acceptable" level of remaining flood risk *after* certain water control works are constructed and nonstructural measures are implemented for Sacramento and Natomas. There were the factors that led SAFCA to choose a project that yielded a level of flood risk greater than that in the NED plan, but that had a lower cost.

The committee is not alone in calling for more attention to residual flood risk. The 1994 report *Sharing the Challenge* also recommended that new floodplain occupants be required to purchase actuarially sound insurance equal to the remaining expected flood damages with the alternative in place (especially for areas that rely on levees, as in Sacramento—see Chapter 5). Such insurance is available through the National Flood Insurance Program (NFIP). The NFIP was developed to ensure that new occupants of flood-prone areas bear a reasonable share of the cost of floodplain occupancy. Expanding the purchase of natural hazards insurance was an objective of a recent bipartisan congressional task force on natural disasters (U.S. Congress, 1994). The task force felt that too-generous disaster aid was an impediment to insurance sale. Its December 1994 report to Congress, motivated by the escalating claims on federal funds by disaster aid payments, stated, ". . . Federal disaster assistance can discourage individuals, communities and state governments from taking action to prepare for, respond to and recover from disasters." The task force went on to state, ". . . if homeowners mistakenly believe that the federal government will rebuild their homes after a natural disaster, they have less incentive to buy all-hazard insurance for their homes."

Requiring floodplain communities and individuals to bear the costs of their hazardous locations will help to inform them of flood risks. At present, the NFIP focuses the attention of its purchase requirements on the 100-year floodplain. By not requiring insurance against residual levels of flood risk, the NFIP may be inadvertently stimulating floodplain development and fostering misunderstanding of flood risk among the general public. In the committee's view, the decision of SAFCA to choose a reduced-size flood risk reduction project, as well as decisions to develop Natomas, has been made in the absence of a requirement to bear reasonable responsibility for the remaining flood risk. Instead, there may be an implicit assumption that such costs will be borne by others, either as disaster aid payments or by a significant federal contribution to construction of flood control works.

In addition to requiring the purchase of actuarially sound flood insurance against residual risk for new development, other land use, emergency planning and floodproofing measures should be pursued for the region as a whole. The first component of a nonstructural plan should be the development and release of information to the public on the nature and extent of flood risks in the region. An inventory of high, medium, and low risk levees should be made available. Developed areas subject to seeps and boils or at some risk from levee overtopping should be designated. Estimates of property damages and life loss owing to dam failure or levee failures in different locations should be provided. Higher risk areas can be compared to lower risk ares and this information can be used to design nonstructural responses for both risk avoidance or to reduce residual risk.

Land use development options near Sacramento can be ranked from lowest to highest in terms of flood hazard risk. For example, a land use development scenario for the Rio Linda area could be compared to a land use development scenario for the Natomas area in respect to relative flood hazards and the risk exposure of new populations.

Regional emergency evacuation plans in the event of dam or levee failures, or levee overtopping, should be prepared and distributed to neighborhood associations and through the media. Residual risks may be reduced by home or business owner actions through temporary floodproofing strategies such as window or door dam placement during high risk, levee-associated episodes. Areas where this is a reasonable options should be identified, so that individual property owners can add their own actions to government emergency responses. A comparable existing model to this in California is property owner participation in earthquake "proofing" of structures and neighborhood planning for earthquake emergency response.

Structural and nonstructural measures should be integrated in the regional response to flood risks and measures can be phased in over time to increase public acceptance and funding opportunities. For example, one integrated package might include reoperating Folsom Reservoir, increasing the capacity of the Yolo Bypass, incremental rebuilding lower American River levees and restoring riparian environments, instituting a flood warning and floodproofing program, and requiring elevation of new structures. These measures could be phased in during the period it takes to finish planning, funding, and building a reservoir for added protection or they could be adopted as the main flood risk reduction measures, if supported by the current USACE analysis. Funding might come from a combination of sources including state environmental restoration programs, federal housing programs, flood control district assessments, and other federal, state, and local sources.

WATER PROJECT COST SHARING

A product of the history of federal involvement in flood control project construction in the American River basin, and the federal provision of flood insurance and disaster aid, is a perception that flood risk management is a federal responsibility. For example, the Natomas development plan is to intensely develop a flood-prone area once an upgraded agricultural levee system is certified by FEMA as providing "100-year protection" under the NFIP. The exposure to life and property from storms of greater magnitude (lower frequency) is recognized, but the presumption seems to be that this remaining flood risk can be ignored, or else that federal or state funds will be employed to upgrade levees and build water control structures.

At the turn of this century, floodwater control was motivated by a desire to reclaim flood-prone lands for economic development purposes. Flood control projects, typically levees, were planned by nonfederal governments and landowner cooperatives who based the desired level of protection on their best technical assessment of the flood risk for an area in relation to the expected values of the reclaimed land. The value of flood protection was established when the funds for project construction were paid by benefiting landowners. As discussed in Chapter 1, the levee districts along the American and Sacramento rivers originated at this time and were established with this financing and planning logic.

Although a full-fledged federal role in flood control was not established until the Flood Control Act of 1936, federal spending for flood protection in the Sacramento area began in 1917. Among the effects of the early federal presence were a better integration of the disparate system of levees and bypasses, the application of analytical efforts that took a basin perspective, and the shifting of a share of the financial burden for flood control from benefiting landowners in the basin to the national taxpayer. The federal financial role was justified by the beliefs that the benefits from flood protection works extended far beyond the lands protected to the nation as a whole and that the costs of floodwater control would stress the ability of local communities to pay for their own protection (Rosen and Reuss, 1988).

Over time the financial responsibilities for flood control project construction have been modified. The 1986 Water Resources Development Act established that for new projects the federal government is to pay between 25 and 50 percent of the cost of construction. Nonfederal interests are responsible for providing all lands, easements, and rights of way necessary for the project construction. These in-kind contributions can be used to offset the required cost share.

The nonfederal cost share for the selected American River dry dam project was estimated to be $240.5 million. Of this cost, $107 million was in the form of lands, easements, and rights of way, primarily for relocation of a bridge on California Highway 49. This bridge relocation was deemed necessary because, when all the storage is employed under an infrequent storm event, road access across the upper canyon area would be cut off for several days. The Sacramento District secured a commitment that the state would make the relocation, but there was no obligation that the relocation be implemented as a condition of the federal project being constructed. Thus, under the cost sharing rules the total cost of $240.5 million significantly overstates the immediate financial obligation of the city and the state. The committee is unable to comment on the cost sharing responsibilities that would arise for the levees or for other alternatives because the application of the formula to consider both in-kind and cash contributions makes such calculations complex. Because of the complexity, the Sacramento District did not report cost sharing burdens for the different alternatives in the 1994 Alternatives Report.

The committee understands the logic behind the original federal financial participation in flood control works—widespread benefits and limited ability to pay. However, it finds that benefits for any American River project are not widespread and that SAFCA, by national standards, has a significant ability to pay. To reach this conclusion the committee first accepts the Sacramento District's analysis that (1) $37 billion of property to be protected is located entirely within the city and nearby areas and (2) that the damages avoided for all alternatives justify the costs (1994 Alternatives Report). Next it calculates the tax burden on the SAFCA community if the selected project had to be paid for entirely by local beneficiaries. Because the available cost estimates are preliminary, these calculations should be considered illustrative. Assume a bond of $600 million for a project sold at 8 percent interest for a 15-year term. At these terms the annual cost to the locality would be around $70 million. This being so, an assessment of $1.75 per $1,000 of the approximately $40 billion of property value would be adequate to repay the bond. For a $200,000 property, the $600 million project would raise property taxes by $400 per year for 15 years.

It is also possible to spread the cost of a protection project over all residents of the city, rather than limiting the burden to flood-prone property. Indeed, there is some evidence from other areas that citizens of a city who do not own flood-prone property are willing to tax themselves to help pay for a project to protect flood-prone areas (Shabman and Stephenson, 1992). While the local cost burden to fully fund a flood control alternative appears significant, if a project is built these costs must fall somewhere in the national economy if they are not borne by those who directly benefit from the project.

COMMUNICATION OF FLOOD RISK

The USACE risk and uncertainty analysis procedures (described in Chapter 4) should contribute to understanding the likelihood and consequences of flood events under different risk management alternatives. Unfortunately, USACE has not developed good explanations of its new procedures and the results from their application. The committee reviewed the risk and uncertainty analysis in the 1994 Alternatives Report and found the explanation and presentation of the results to be particularly confusing. No distinction is made between estimates of flood risk calculated with the traditional level of protection methodology and those calculated with the new risk and uncertainty methodology. Both are called "level of protection" and described by a variety of terms. The most common terms in the report are control for T years, T-year level of protection, T-year protection, and T-year flood protection. Yet the use of the term level of protection to describe flood risk is inconsistent with the new USACE guidelines for risk and uncertainty analyses (EC 1105-2-205, 1994) and confuses the traditional and the new approaches to calculating flood risk.

Finding an appropriate way to express the concept of flood recurrence interval and the T-year flood is a persistent problem in risk communication. It seems to be fairly well understood that floods (or, more precisely, exceedances of a specified stream flow rate) can be described as a series of random trials with a probability p of "success" at each trial, and with an average time between successes (recurrence interval) of $1/p$ trials. (See, for example, Interagency Floodplain Management Review Committee, 1994.) Nonetheless, most people do not have much experience with the distribution of times between floods, or much experience in describing or discussing such phenomena, and their experience with more familiar random processes such as seasonal rainfalls and temperatures is not directly transferrable to the time distribution of floods. For phenomena like seasonal rainfall or temperature, the average value also is a fairly typical value; the observations tend to be concentrated in some interval in the neighborhood of the mean, and observed values become progressively less frequent at greater distances above and below this interval. The frequency distribution is somewhat bell-shaped, perhaps with some degree of asymmetry or skewness. The frequency distribution of times between occurrences of floods exceeding a given magnitude, however, does not have this shape. For exceedances of a given flow rate, the frequency distribution of inter-arrival times has a maximum at year 1 and decreases progressively for longer inter-arrival times. For exceedances of the 100-year flood (the flow rate having 1 percent chance of being exceeded in any year), for example, the probability is 1 percent that the time between occurrences will be 1 year, 0.99 percent that it will be 2 years, and, in general, $1 \times (.99)^{(n-1)}$ that it will be n years. There is no tendency for the inter-arrival times to cluster around the mean value; such a tendency would exhibit itself as a periodicity in flood occurrences, and such periodicity is not generally observed upon examination of flood records. Nonetheless, the temptation to think of the mean value as a typical value is strong, and use of the mean time between occurrences (recurrence interval) in risk communication may lead to misinterpretation and misunderstanding.

Efforts have been made to improve flood risk communication by avoiding use of the term recurrence interval and emphasizing the concept of percentage chance of occurrence. It is not known whether these efforts have borne fruit. The concepts and terminology of recurrence interval and *T*-year flood are deeply rooted in the vocabulary of land planning, development, engineering, insurance, and regulation. The 100-year floodplain is the cornerstone of the National Flood Insurance Program and all its associated regulatory and management apparatus. It seems unlikely that these concepts and terms can be eliminated and, considering that they are valid and legitimate concepts, it is not clear that they should be eliminated. It appears that risk communication might best be improved not by use of a limited vocabulary but by improved explanation of the hydrologic and statistical concepts underlying flood risk assessment. One promising way of accomplishing this is by re-casting probability statements about flood risk into terms with which citizens have direct experience. For example, an annual probability of flood occurrence of 1 percent is not readily comprehensible to most nonspecialists; the equivalent statements of more than 1 chance in 6 over a 20-year period or more than 1 chance in 4 over the life of a 30-year mortgage may be more readily interpreted. Also, comparisons with other forms of risk can be valuable.

Risk communication is required in many areas of environmental policy. These include regulation of air and water pollution, regulation of hazardous waste and chemicals, issuance of licenses for pesticide use, and review of drugs and the safety of food and food additives. Federal agencies have developed a risk analysis/risk management framework (NRC, 1983; Cohrssen and Covello, 1989) that includes risk communication as an important component of risk management (NRC, 1989; Plough and Krimsky, 1987). The new USACE risk analysis procedures allow the transfer of this framework to flood risk management. This transfer should occur. Public officials, the general public, and home owners need to understand the magnitude and consequences of floods if they are to make informed risk management decisions (Sandman, 1985). "It also allows creation of the trust and understanding necessary for successful risk management within our political system" (Slovic, 1987). The two dimensions of flood risk—likelihood of an event and its consequences—need to be communicated.

The most commonly used description of flood risk is the chance of flooding in a single year. This is often represented by a "return period", which has been called the "level of protection." Level of protection highlights neither the negative consequences of a flood event, its probabilistic nature, nor the differences in the confidence that can be placed in different alternatives to achieve a similar calculated level of protection. USACE has committed itself to avoiding the term level of protection, but continued to use it in the 1994 Alternatives Report.

"Chance of flooding" is probably a better description of flood risk than is a term like level of protection. Phrases like chance of flooding direct attention to the hazardous event of concern (see Chapter 4). Using chance of flooding also avoids use of return period. Return period has been incorrectly understood to mean that one and only one T-year event should occur every T years. For example, an article in the Sacramento Bee (Hicks and Blechman, 1994) describing the 1994 Alternatives Report stated that, "Currently, Sacramento levees could withstand a flood of such magnitude that it is expected to occur only once every 78 years." Unfortunately, this misleading statement was extracted from page 1 of the 1994 Alternatives Report.

With the new USACE risk analysis procedure that was employed, failure is no longer related to a single T-year design flood, but can depend on different combinations of flood flows, operating decisions, and levee performance. The risk analysis procedures recognize that the flood control system might not perform its expected flood control function. For example, levees may not be 100 percent reliable even at design flows, and so the likelihood of flooding is determined by more than the rainfall and runoff frequency.

In the American River case, the factors that can affect the performance of both reservoirs and levees must be evaluated. The most common form of risk assessment involves identifying events or a sequence of events that can relate to dam or levee failure, identifying the modes of failure, evaluating the likelihood of a particular mode of failure, determining the consequences for each potential failure mode, and calculating the risk costs or expected economic and social losses from the failures. Risk assessments for dams would consider the potential for and consequences of overtopping, flow erosion, slope protection damage, embankment leakage and piping, sliding, deformation, deterioration, earthquake instability, construction problems, gate failures, erosion of spillway, operator error, and catastrophic failures (NRC, 1985). Risk assessments for levees would consider similar factors, such as overtopping, flow erosion, slope protection damage, embankment leakage and piping, subsidence, earthquake instability, construction problems, operator error and catastrophic failures.

The public also must recognize that flood risk has another dimension—consequences. Levee building can provide a good illustration of the kinds of consequences that should be discussed and evaluated in association with a flood risk reduction project. In rivers not bounded by levees, rises in stage usually most affect structures nearest the river, with the effects of overbank flow diminishing with distance from the river. In other locations where levees protect structures in the floodplain, when flows overtop the levee there is a discontinuity in the relationship between flood flow and stage, and its impact on floodplain structures and inhabitants. When a levee fails, the sudden rush of water may inundate structures by many feet of water and cause more damage than would have occurred if the levee had not been constructed. A levee failure is akin to a flash flood. Moreover, the development that might be encouraged because of the partial protection levees provide can translate into greatly increased damages if a levee fails.

When a reservoir fails to contain a flood, only the extra flood volume is spilled, so that flood stages may rise only a little above the target maximum flood stage. Thus the initial reservoir storage is not lost. By contrast, when a levee is overtopped or breached, it loses all its effectiveness.

The formal evaluation of flood control benefits is one reflection of the likelihood and consequences of flooding. When real property is inundated, the expected annual damages reduced by the project are the NED benefits. Although this dry economic calculation may help to select from among alternatives, it cannot be the only basis, because NED benefits do not convey the consequences of extreme events. First, extreme events have a low probability and so are weighted quite low in the expected damages calculation. Second, the focus only on real property losses may not reflect the different duration of flooding or the rapidity of the rising waters and the threat to evacuation opportunities. Table 6.1 gives several other representations of flood risk that can be used to improve risk communication.

Even the numbers that might be developed from applying this table may not adequately convey flood risk in places like Natomas and Sacramento. Reliance on levees of uncertain structural integrity introduces a factor that may not be adequately reflected in Table 6.1. When projects must be operated at a high degree of technical efficiency, there is a hidden possibility of operator failure.

TABLE 6.1 Aspects of Flood Risk and Possible Descriptions

Safety	Annual probability of flooding
	Chance of flooding during a structure's economic life
	Chance of flooding during 20-year or 50-year period
	Expected loss of life should a flood occur
	Expected injuries
Physical severity of floods	Depth of flooding in different areas
	Duration of flooding
	Velocity of water
	Areal extent of flooding
	Physical damage to flooded areas (effect of sediment on soil, soil loss, effect on vegetation, etc.)
	Length of time before economic activities could resume
Economic damages	Value of property at risk
	Dollar loss that would occur with a particular event
	Expected annual damages due to flooding
	Expected annual damages for a residential home
	Annual insurance costs for a policy on flood damages for a home
	Cost of lost business due to disruption from flooding

The character of the area that would be inundated is also an important consideration. Floodplains surrounded by rivers and having a bowl shape present a different hazard than land that slopes up as one moves away from the river. How deep would the water be? How much warning would there be? Could people escape?

Some of the statistics described above could help illustrate these issues. One way to add diversity to descriptions of flood risk is to have USACE, *in cooperation with those in the decision process,* create realistic risk scenarios for the different alternatives. The cooperative building of scenarios can be an excellent way to communicate flood risk. For example, the following scenario is one possible description of the vulnerability of the Sacramento and Natomas areas for storm events that overtop the levee system:

> Should levees protecting Sacramento south of the American River be threatened, residents could attempt to move to higher ground to the south and west farther away from the river, and the depth of flooding would generally not exceed that at the rivers edge; few areas would experience flooding of more than 10 feet. Natomas, on the other hand, is ringed by levees so that residents trying to leave the area would have to find their way across the main highway system to areas with higher ground that are primarily to the west. Moreover, because Natomas is in a depression, a third of the area would flood to over 10 feet, and some to as much as 35 feet in depth. If the Natomas area is subject to a 1 in 100 chance of being flooded in any year, then the probability of at least one flood in 50 years is 40 percent. Therefore, the probability of a relatively catastrophic event within the lifetime of most residents is roughly equal to the probability of flipping a fair coin and getting heads.

IMPROVED APPROACHES TO
FLOOD RISK MANAGEMENT PLANNING

As a result of laws such as the National Environmental Policy Act (NEPA) and the increased power of stakeholders such as environmental groups, the USACE planning process has allowed increased public participation in all stages of planning. The American River planning process, for instance, included three public hearings, receipt of more than 2,000 comment letters and more than 650 pages of response by USACE (1991 ARWI, Appendix T). The result has been escalating conflict over major water management alternatives. Conflicts can occur when people disagree over "facts." People can also disagree because they feel their interests are not being equitably served and because they have different values (Lord, 1979). In the American River case, value conflicts are especially sharp. For example, as stated in one recent newspaper article, ". . . you have two warring sides: environmentalists and dam supporters. They're religious wars; conflicts of values that are unresolvable" (Hicks and Blechman, 1994).

The ability of USACE to act unilaterally in resolving conflict has diminished over the past three decades. In the past, there was a national consensus behind water development, the public accepted without question the expertise housed in executive agencies, and access to the courts and to the legislature to oppose agency decisions was more limited. Today, decisionmaking on any plan requires agreement among those who are affected by, and can block or advance, implementation of an alternative in a variety of different legal and political forums. Resolution of conflict among those who can affect a decision could perhaps be facilitated by the USACE planning process, although it is unlikely that USACE planning and budgeting processes alone will be adequate for reaching all necessary agreements for implementation of most large water projects.

Plan formulation demands the creation of the widest possible range of engineering and institutional alternatives so that agreements can be reached among multiple decisionmakers. The way that plans are formulated can secure support from affected interests. Failure by the Sacramento District and SAFCA to initially incorporate a wide range of purposes and institutional adjustments as a part of plan formulation, and to open the planning process to multiple interests, has been a barrier to agreement on a flood risk management alternative for the American River.

Federal flood control planning in California in the past decade and a half has been characterized by citizen groups or local governments who are not project sponsors hiring their own consultants to determine what alternatives are technically feasible and to describe the social and environmental impacts mentioned in federal reports more fully. Cases where this has occurred in federal projects in California include not only the ARWI for Sacramento, but also Tecolote Creek, San Diego; Mission Creek, Santa Barbara; Soquel Creek and San Lorenzo River, Santa Cruz; San Pedro Creek, Pacifica; Walnut Creek in the City of Walnut Creek; Wildcat Creek, Richmond; Murderers and Grayson Creek, Pleasanton; Corte Madera Creek, Ross; Napa River, City of Napa; Dry Creek, Roseville; and, Petaluma River, Petaluma (E. Cummings, California Department of Water Resources, personal communication, October 3, 1994).

In the American River case, a variety of pro bono consultants were brought in to help the Planning and Conservation League evaluate the draft environmental impact report and propose project alternatives not provided by the USACE reports. The Environmental Defense Fund hired a hydraulic engineering firm to propose alternatives (Environmental Defense Fund, 1990; Jennings, 1991). The ad hoc planning investigations suggested alternative approaches to lowering flood risks, including redesign and reoperation of Folsom Dam, operating upstream hydroelectric reservoirs for flood storage, use of surcharge storage in Folsom, and the potential for greater storage in the Sacramento River bypass system.

Since the congressional action on the 1991 ARWI, SAFCA has hired consultants in geomorphology to help it develop a plan to integrate the rebuilding of lower American River levees with riparian habitat restoration. This work has been an analytical effort in support of the Lower American River Task Force, formed by SAFCA to build a consensus among general public and agency stake-holders on finding compatibility between environmental restoration and flood control purposes on the river below Folsom (SAFCA, 1994a). The original proposal to riprap 20 miles of river levees along the American River has been reframed to integrate levee strengthening and riparian restoration projects. By-pass expansion, levee setback, and Sacramento River riparian restoration alternatives are being studied. On the upper river, SAFCA has pursued research to help in understanding potential ecological impacts of a dry or multipurpose dam.

Meanwhile, an interagency Yolo Basin Working Group and USACE began an assessment of how to integrate flood protection and environmental restoration purposes in the Yolo Bypass. By 1994, interagency agreements had been approved for the multi-objective management of the bypass for endangered species protection, wetland and wildlife habitat restoration, and flood control (California Resources Agency, 1993; Yolo Basin Foundation, 1994). The intent is that the lower American River levee and bypass improvement projects will enhance the American River Parkway and bypass aesthetic and environmental values while also performing a flood control function.

The American River experience prior to 1991, with its focus on flood protection through new storage, resulted in a planning process that was unable to define acceptable alternatives. Failure to expand the planning purposes meant that the Sacramento District could not develop a broad constituency of support, because the plan made flood control a competing purpose against habitat restoration and water supply, rather than developing an integrated package. If multiple purposes had remained the focus of planning, as had been suggested by the original study authorities, alternatives capable of coalition building might have emerged. For example, the failure to include environmental restoration in the lower river as a planning purpose, and the treatment of environmental considerations only as mitigation requirements for the dry dam, may explain the difficulty in reaching agreements. In the American River situation a restoration planning purpose would have focused immediate attention on riparian habitat in the design of levee repairs and on water flows to the delta in the formation of operational rules for all projects in the basin.

However, two shifts in the USACE planning process have occurred, each of which should create an opportunity for a new approach to plan formulation. USACE now encourages a broader conception of project planning purposes, including environmental restoration, although change has been difficult (Riley, 1989). Also, USACE has emphasized planning partnerships in many controversial water management situations (e.g., Columbia River and Florida Everglades (Shabman, 1993)). For USACE planning to meet the "acceptability" test of the P&G in a planning partnership, the agency must understand the barriers to agreement and then suggest and analyze institutional and operational, as well as engineering, measures in plan formulation as a means of securing agreements.

Resolution of the risk management disagreements for the American River might be advanced by this new planning approach. A necessary step is to fully incorporate water supply management, recreation, environmental restoration, and hydroelectric power purposes throughout the planning process. The Sacramento District appears to have directed its attention primarily to flood risk reduction and in so doing has motivated opposition to its recommended flood control alternative. This planning approach might be explained by the original authorization language for the study (stressing flood control and a dry dam), or it might be explained by USACE budget priorities in the 1980s being limited to flood control and navigation (Shabman, 1993). Another possibility is that the district saw its role in meeting these purposes strictly through construction of a storage project or levees. If this was the case, then attention to these purposes, if water project construction alternatives were the solution, led to a full-pool Auburn reservoir, and the district concluded that this was not a viable alternative. In the 1991 ARWI the closest the Sacramento District came to addressing multiple purposes was to discuss the option of converting the dry dam to a full pool at some future date.

However, alternatives are not simply engineering measures. They are also the institutional agreements that determine the financial, legal, and political acceptability of a project. Institutional adjustments that would be required for an aggressive reoperation of Folsom, for use of upstream storage, for levee upgrades, or for building acceptance for a dry dam have not been made part of the plan formulation process. The committee has found two areas where this opportunity may have been missed, but emphasizes that the discussion of these opportunities is not meant to endorse any flood risk management approach.

One illustration can be drawn from the expressed concerns about purposes other than flood control. While reoperating Folsom under conditions when the reservoir does not recover its storage might reduce the reliability of the water supply, technical means might be available to ensure reliable water from other sources, but these have not been explored. For example, water released to create flood control storage might be retained underground in overdrafted ground water basins in the southern part of the state (Jaguette, 1978). Another option to meet water supply contracts would be to pump ground water from the Sacramento groundwater basin, which currently has relatively low lifts (and hence costs) in some parts of the basin. Protection against long-term ground water overdraft and compensation to local ground water pumpers would be a part of any such alternative. Representatives of the Central Valley Project's (CVP) power contractors have similar concerns about the availability and costs to their users from Folsom reoperation. Also, it could be very difficult to offset losses to recreational stakeholders, and it may also be much more difficult to meet delta environmental requirements during drought years without adequate water in Folsom. However, the use of Folsom will not result in massive losses to these interests in every year, only in years where flood control operations are driven by large storm events.

Perhaps some arrangement might be made in consideration of the low frequency of the effects.

Additional study would be required to determine if these are technically workable measures; however, the main point is that they all would require institutional adjustments through water and power markets and in other ways as well. Thus if flood control beneficiaries of Folsom reoperation are willing to pay to purchase pumped ground water and alternative power to make up for any losses to the CVP, opposition to Folsom reoperation might be reduced. The Sacramento District or its planning partners in the state and in SAFCA need to incorporate analysis of such institutional measures in any plan formulation that includes reoperation. As this report was in final preparation, SAFCA entered into a contract with the Bureau of Reclamation that appears to implement many of the institutional measures recommended here for gaining agreements on the reoperation of Folsom (U.S. Department of the Interior, 1994).

Successful planning will recognize the interests that sought a full-pool Auburn reservoir and will define measures that might bring them into an agreement. These measures will probably not center on obtaining "wet water" but on facilitating water transfer agreements, cash payments for water conservation programs, or support for construction of water pumping and transfer facilities. While these measures fall outside the Sacramento District's view of its implementation authority, the failure to include such measures in plan formulation means that a source of support for any flood risk management alternative is lost.

At some point, it may be determined that a desired amount of risk reduction will require a dry dam near the Auburn site. Opponents of the dry dam are concerned about two possible outcomes of such a decision, but institutional measures might be developed to facilitate agreement over concerns about impacts to the American River canyon. First, opponents believe that a dry dam will eventually be converted to a full pool. It is considered politically prudent to oppose any dam, lest a dry dam be the first step toward a full-pool reservoir. SAFCA's suggestion for addressing this concern was to propose an ungated structure. The committee finds an ungated design to be imprudent (see Chapter 3). Earlier in this section, the committee suggested the kinds of institutional measures that can address the water supply purpose and in so doing reduce the pressure to construct a reservoir with a permanent pool. In combination with such institutional measures, it might also be possible to ensure the pending wild and scenic status for the American River, or to transfer canyon land ownership to a natural area trust, as a condition of dry dam construction.

A second concern voiced by opponents to the dry dam is about the uncertain environmental effects of occasional canyon inundation. The Sacramento District has stated that inundation of the canyon would occur several years out of ten, although the exact frequency has never been calculated because the district apparently has not completed and reported on the flood control system operation procedures for the dry dam. All that the District has suggested to the committee is that the projects will be operated jointly, meaning that for high-frequency events there will be some canyon inundation.

The committee believes that a suggestion of the Planning and Conservation League of California with respect to the dry dam (Jennings, 1992) has received too little attention as a possible focus for agreement (if there is a concern that the risk of a no-storage option is unacceptable). The suggested operational strategy would be to use the dry dam as "insurance" against extreme flows, driving down the likelihood of use of the dry dam (for example, it might impound water only in the 100 year—1 percent chance—event). The first line of flood protection would achieve high levels of flood damage reduction from a modified and reoperated system with Folsom as the key element. What is significant here is that this approach would reduce the likelihood of the dry dam impounding water, and hence the possibility and frequency of environmental impacts occurring in the canyon would be reduced. This strategy would require gates for operation of the dry dam. As a bonus, a radically reduced frequency of inundation may make relocation of highway 49 less necessary, reducing total project cost by over $100 million.

Also, a mitigation strategy for the uncertain effects of infrequent inundation would be needed. An institutional response would be the creation and funding of an "adaptive management" trust fund for the canyon. This fund would have adequate assets to compensate for restoration or replacement elsewhere of the environmental assets of the canyon following any inundation. It would be initially endowed by the beneficiaries of the flood control. Withdrawals from the fund would be made after each inundation, and the amount withdrawn would be replaced by charges against downstream flood control beneficiaries. With these charges as a consideration, the use of the canyon as a last element used in any flood event would be encouraged.

The committee recognizes that creativity in identifying and designing institutional measures may fall outside the authority of the Sacramento District. This being the case, and given the central importance of institutional analysis, USACE policy may need to encourage the Sacramento District to structure its planning process differently. The environmental impact statement (EIS) established by the National Environmental Policy Act of 1969 and the Fish and Wildlife Service review of projects under the Fish and Wildlife Coordination Act of 1958 provide for public and agency participation in the formulation of USACE plans. The development of the EIS, as practiced by the District, limits the public and other local, state, and federal agencies to project review roles rather than integrating their contributions into plan formulation. Early integration of disparate interests could focus on building consensus about a full range of measures to be considered, including those that might lie outside the engineering and construction mission of USACE. However, there is no assurance that other agencies and interests will bring the necessary creativity to a newly opened planning process. There is no formula for selecting any mix of agencies and public organizations that will bring these insights to planning, but simply acknowledging the need for such creativity is a step in the necessary direction.

THE WATER POLICY AND MANAGEMENT CONTEXT

Flood risk management decisions for the American River must be addressed within the larger context of California water management. Certain recent federal environmental rulings may affect the opportunities to implement some of the flood risk management alternatives for the American River. Oversight and coordination of studies and proposals related to flood control and to other aspects of water management will be required if any alternatives to address the potentially conflicting purposes are to be implemented. The Sacramento District should not be expected to, nor is it able to, provide this oversight and coordination function.

There are at least three relevant major federal proposals in various stages of planning and implementation that relate to these environmental concerns: (1) a proposal by the Environmental Protection Agency (EPA) to increase water flow to the Sacramento Delta to reduce the salinity intrusion and its effect on biota, (2) a proposal by Fish and Wildlife Service under the Endangered Species Act to increase flows to the delta to benefit the delta smelt and to declare threatened status for the Sacramento splittail, and (3) several provisions of the Central Valley Project Improvement Act (P.L. 102-575, Title 34). There are also several related state proposals.

Recently, EPA proposed a rule establishing three different water quality criteria for the delta: (1) salinity criteria protecting estuarine habitat in the Suisun Bay area, (2) salmon smolt survival indices protecting salmon migration, and (3) an electrical conductivity criterion protecting striped bass spawning on the lower San Joaquin River (EPA, 1994). The primary method for implementing these proposals involves increases in delta outflow. EPA estimated that the increases would average 540,000 acre-feet per year and go as high as 1.1 million acre-feet in very dry years. These increased outflows would be needed primarily from February through June.

In March 1994 the California State Water Resources Control Board disputed the proposed criteria (CSWRCB, 1994), and EPA subsequently revised its proposal. In an effort to resolve this dispute between federal and state regulatory agencies, which had been ongoing since 1978, then Assistant Secretary of the Interior Elizabeth Rieke and California Secretary for Resources Wheeler announced in June 1994 a Framework Agreement between the Governor's Water Policy Council and the Federal Ecosystem Directorate. The agreement is designed to resolve bay-delta management issues and coordinate the regulatory process. Federal agencies involved in the pact are the National Marine Fisheries Service, Bureau of Reclamation, Fish and Wildlife Service, and EPA. State participants are the California Resources Agency (Department of Fish and Game,

Department of Water Resources) and the California Environmental Protection Agency (including the State Water Resources Control Board).

A major reconciliation was announced on December 15, 1994. State and federal officials, in cooperation with urban, agricultural, and environmental interests, agreed to implement a specific plan to provide ecosystem protection for the bay-delta estuary. The provisions are intended to be in place for 3 years, at which time they would be reviewed for possible revision. They include water quality standards, limitations on exports from the delta, and assurances that the plan is intended to create conditions in the delta that will avoid the need for any additional listings under the Endangered Species Act during the next 3 years.

It will be some time before final decisions are made implementing this proposal and before it will be known how the additional flow, if mandated, will be allocated among the various tributaries. However, it is almost certain that Folsom operations would be affected by any increased flow requirement. Owing to its proximity to the delta, Folsom is the first in line to ameliorate adverse water quality in the delta.

In January 1994 the Fish and Wildlife Service released two related documents (USFWS, 1994). One proposes threatened status for the Sacramento splittail (a large cyprinid fish); the other proposes the delta area as critical habitat for the delta smelt. These two proposals are closely linked to the EPA salinity criteria mentioned above, and the two agencies are working together to present a unified program to benefit the relevant species.

The Central Valley Project Improvement Act (CVPIA, P.L. 102-575, Title 34) includes two major sections that may affect water routing from the American River (USDI, 1993). The first is an ambitious plan to double the numbers of anadromous salmonids in the Sacramento-San Joaquin basin by the year 2002, compared to the base period of 1967 to 1991. It would involve all of the major tributaries of the Sacramento. The planning process is in very preliminary stages, and there is little documentation available at this time. The two primary requirements are temperature control and increased minimum flows.

The doubling plan for anadromous fish in CVPIA is consistent with one established by the California legislature in 1988 (Chapter 1545/88). In the implementation of the state plan, the California Department of Fish and Game has a proposal for anadromous fish enhancement (CDFG, 1993) that presumably will be integrated into the CVPIA program. The other relevant section of CVPIA directs the Secretary of the Interior to "dedicate and manage annually 800,000 acre-feet of Central Valley Project yield for the primary purpose of implementing fish, wildlife, and habitat restoration." The source of this additional water is not specified and is the subject of much debate as the new law is being implemented. In the first year, part of the 800,000 acre-feet was dedicated to maintaining releases of no less than 1,750 cubic feet per second from Nimbus Dam from October 1993 through February 1994 (USDI, 1993). Water provided from CVP for delta water quality under the December 1994 state-federal agreement will be credited to the 800,000 acre-feet of obligation.

In the continuing effort to solve the myriad problems caused by too much or too little water in the American River basin, more may be asked of Folsom than it can possibly deliver. Efforts to develop reoperation plans will have to take these demands into consideration. Coordination with other environmental concerns such as the CVPIA needs to take place to establish regional ecological management priorities. What is needed is enhanced capability for technical systems analysis to support decisionmaking on conflicting water resources goals. Meanwhile, the state's dedication of resources to data collection and statewide planning that might resolve conflict has been radically reduced in recent years. Further, California's Department of Water Resources is now largely funded by revenues from state water project contractors, who represent mainly urban and agricultural water supply interests. Limited resources to conduct data collection, to conduct water policy research, and to create conflict resolution mechanisms may inhibit the ability to reach agreements on such matters as competing demands for the storage at Folsom.

CONCLUSION

Agreements on acceptable level of flood risk and on the alternatives to achieve that level have escaped the American River planning process. If there is to be a resolution of the issue of appropriate flood risk for the American River, alternative planning approaches and new leadership from outside the Sacramento District, while drawing on the resources and expertise housed in USACE and the district, will be needed. Specific attention must be paid to enhanced risk communication to ensure that the full costs and benefits of different alternatives are adequately understood by the public. Toward this end, federal policy should increase the cost responsibility on project beneficiaries.

Based on careful consideration of these issues, the committee believes that the following recommendations should be considered:

• Future federal participation in flood damage reduction projects for the American River should be conditioned upon SAFCA, working with FEMA and private insurers, requiring landowners to purchase actuarially sound flood insurance against residual risk for new development at Natomas and for Sacramento. In particular, SAFCA should either (1) institute a program to require that individuals purchase their own flood insurance related to the risk of the location and development or (2) purchase group insurance for all landowners in the region and recover purchase costs by assessments on landowners who receive coverage.

• USACE, FEMA, and other federal agencies should adopt an agreement governing federal participation in structural and nonstructural flood risk management efforts to require that benefiting local communities have a program requiring new development to purchase flood insurance at actuarially sound rates for residual flood risk. Existing development should also purchase residual risk insurance, presumably at lower rates than new development. The federal government, working with private insurers, should develop provisions for sharing the cost of flood insurance premiums with communities and individuals who implement structural and nonstructural flood damage reduction measures.

• Before authorizing additional federal financial commitments for flood control on the American River, Congress should explicitly determine whether flood control on the American River constitutes a problem warranting federal involvement based on the presence of widespread national benefits or the limited ability of the community to provide for its own flood protection.

• Congress should reform cost sharing requirements in the 1986 Water Resources Development Act (1) to increase the nonfederal cost share significantly above currently authorized levels, granting exemptions to the higher rate when it is demonstrated that flood control benefits are widespread or that the benefiting communities have limited ability to pay for justified flood protection and (2) to first define all cost sharing requirements as a share of total project costs and then make allowances for documented in-kind contributions to be counted toward the allocated cost share.

• The Sacramento District and SAFCA should report the flood risk on the American River as a chance of flooding of 1 in 100 per year (or whatever figure is appropriate). Such annual risk figures should also be converted to the risk over longer time periods. For example, a 1 in 100 year risk results in a 40 percent chance over the next 50 years that floodwaters will overtop or breach the levees and inundate the Sacramento area. The 1986 flood in the area, which has been estimated as a 70-year flood event, should be used as a reference to convey the magnitude of larger and less frequent storms.

• The Sacramento District should act with SAFCA and other stakeholders to build and publicize realistic scenarios to describe the consequences of a levee being overtopped. Descriptions of the vulnerability of the Sacramento and Natomas areas to storm events that overtop or breach the levee system should clearly address the extreme depth of flooding possible, the transportation difficulties that will be faced, and the problems involved in recovering from flooding in a closed basin.

• The traditional term "level of protection" misleads the public and is not consistent with the analytical outcomes expected from the new USACE risk analysis procedures. Therefore, USACE should select a technically sound risk communication vocabulary and approach to communicating flood risk likelihood and consequences (see Chapters 4 and 5) and use it consistently in all its reports and presentations. In addition, USACE should work with FEMA, as well as other agencies, the states, and private insurers, to develop a standardized vocabulary that adequately conveys flood risk and vulnerability.

• The Sacramento District and SAFCA should expand the consensus-building efforts of the Lower American River Task Force for the purposes of (1) addressing the full array of purposes that were originally part of the study authorization, including water supply and allied purposes and (2) identifying institutional agreements that can employed to address these purposes. In this effort, the district and SAFCA might request the leadership and assistance of the State of California's Resources Agency.

• USACE should issue guidance to its districts stressing the requirements to maintain a broader view of water resource planning purposes and to address those purposes throughout the planning process in the development and evaluation of institutional as well as engineering measures.

• The state of California should not expect the flood control controversy on the American River to be resolved solely through federal leadership. The state needs to increase its participation in, and the resources it dedicates to, the basinwide water resources planning needed to build a consensus on technical and institutional strategies to manage competing water demands, including flood risk management.

• For especially contentious disagreements, USACE should advise its Districts to facilitate but not dominate the local decisionmaking process. That role includes provision of technical analysis as well as initiation of and participation in decisionmaking forums (such as that initiated by SAFCA in the lower American River). USACE districts should assist agencies of the federal and state governments and nongovernmental interests to cooperatively develop the data and models, understanding of risks and trade-offs, and possibilities in the formulation of alternatives early in the planning process.

7

Findings and Recommendations

The Committee on Flood Control Alternatives in the American River Basin was charged to review technical and policy issues that arose from the Sacramento District's 1991 American River Watershed Investigation (ARWI) and, where possible, subsequent planning activities. The committee's task was to evaluate the scientific and engineering knowledge on which the selection of a flood hazard reduction strategy for the area will ultimately be based. The committee also endeavored to provide insights on public policies concerning flood hazard management that are of concern to the nation.

In line with that dual charge, the committee offers some thoughts specific to the USACE planning process as it was applied to the American River basin. The committee also comments more broadly on the nature of flood risk assessment and its application nationwide. The findings relate to (1) the identification and evaluation of alternatives in Sacramento District planning documents, (2) environmental issues in the upper American River basin, (3) risk methodology, (4) flood risk management behind levees, and (5) the implications of the American River example to resources planning and decisionmaking.

The key issue in the planning process, and in this report, is how to reduce flood risk in the lower American River basin given a decisionmaking arena that includes significant scientific uncertainty and organized opposition to some of the possible risk reduction alternatives. This report discusses the uncertainties that confront floodplain managers and offers suggestions in many areas, including the need for additional research. But decisionmakers, agency officials, and interest groups reading this report should not use calls for additional research as an excuse for not taking action.

It is time to select and implement flood risk reduction strategies for the American River basin. There are still areas where data and information are incomplete, particularly in our understanding of environmental impacts, but that should not forestall the decisionmaking process. Data collection and interpretation can continue as the decisionmaking process proceeds. Adaptive management techniques can be used to select approaches that can be monitored, evaluated, and revised as implementation proceeds and additional information becomes available.

It is important to understand that even if Sacramento achieves its stated goal of a "200-year level of protection," the city will still face a significant residual flood risk. The risk would be equal to a probability of flooding of 1 in 200 per year or about 22 percent over a 50-year period, essentially a 1 in 5 chance over 50 years. If flood risk in the city of Sacramento or Natomas is 1 in 100 per year, then the residual risk over 50 years is about 40 percent, or about the probability of getting "heads" with a single flip of coin.

Moreover, estimation of the residual risk of flooding alone does not provide owners and occupants of facilities in the floodplain with a complete picture of the consequences and damages that are likely to result from flooding. Estimates of flood risk should be augmented by estimates of likely loss of life and property damages, which are affected by evacuation opportunities, warning times, and the likely depth and character of flooding. Such vulnerabilities can be communicated by realistic scenarios that illustrate how a flood event would look and what losses are likely to occur.

Perhaps the worst thing that might be done is to create a false sense of security or to encourage people to think that any proposed project provides complete protection from flooding. Therefore, flood risk management needs to be an ongoing part of urban planning in the city of Sacramento, including in particular the Natomas area, to reduce residual vulnerability to disastrous flood losses. One element of such management is improved flood risk communication, which would give investors and residents in the area a better understanding of the risks and vulnerabilities they face.

It is increasingly evident that the nation has entered an era when construction of new water projects will be rare. Budget constraints, environmental considerations, and organized interest groups all contribute to this situation. Because society is less willing to build new facilities, we must find ways to obtain more from existing facilities. We need to operate existing facilities more efficiently and to upgrade planning methods, system instrumentation, modeling capabilities, and other tools used to support operation. With better knowledge, we can make existing systems more efficient and more responsive to public needs without necessarily accepting greater risks. Such improvements would not preclude all new construction, but rather would reduce the need for new facilities.

Despite this general admonition to strive for greater efficiency from existing systems, the committee cautions that reoperation of Folsom Dam and Reservoir cannot solve all the water problems faced in the American River basin. In the struggle to solve the myriad problems caused by too much or too little water in the Sacramento area, more may be asked of Folsom than it is able to deliver.

The issue decisionmakers face is how best to determine and then implement an acceptable flood risk management program for the American River basin. Beyond all the complexities and subtleties, the ultimate question is whether the flood damage reduction offered through a combination of measures not including a new dam is acceptable, or whether a new upstream dam is necessary to reduce risk to a more acceptable level. The committee cannot answer that question, in part because detailed technical analyses comparing the current range of alternatives are still being developed (these analyses are expected in the Sacramento District's forthcoming supplemental information report, scheduled to be available in the summer of 1995) and, importantly, because that judgment is beyond the committee's authority or appropriate role.

The public should be forewarned that even when the technical analyses are available, there will be no simple technical answer. Scientists and engineers can and should provide careful analyses and interpret the information so it is available to support decisionmakers, and they should be frank about uncertainties and risks. But the decision to be made should ultimately reflect more than technical factors; it should reflect economic considerations and value judgments pertaining to the appropriate use of natural resources, public monies, acceptable levels of risk, and willingness to accept constraints on land use. The final decision on these issues rests with the public and the political officials who represent them.

IDENTIFICATION AND EVALUATION OF ALTERNATIVES

Committee's Reaction to the 1991 and 1994 Reports

The structural flood protection measures and alternatives described in the 1991 American River Watershed Investigation (USACE, Sacramento District, 1991), as supplemented by the 1994 Alternatives Report (USACE, Sacramento District, 1994a) are reasonably complete, although supporting analysis for the 1994 document remains to be seen. Alternative assumptions could have been selected, as discussed in Chapter 2, but the committee found no omissions or errors of a degree that should call the overall results into question.

Planning Reoperation of Folsom Dam

One concern identified soon after publication of the 1991 ARWI involved the operating policies assumed for analysis of flood control effectiveness of Folsom Dam. Since then, ongoing investigations of interim Folsom reoperation have been exploring a more dynamic allocation of Folsom storage capacity based on the level of storage in upstream reservoirs. The 1994 Alternatives Report assumed that the reoperation plan will be adopted as part of the without-project alternative. For alternatives that involve new low-level Folsom outlet works, the analysis should include modification of the Folsom operating policies to take full advantage of the new release capacity. Likewise, planning studies of a proposed dam at Auburn should investigate how to operate the dam to achieve the desired flood protection at Sacramento while minimizing environmental impacts. The committee was concerned by the fact that it was unable to evaluate how Folsom reoperation and other alternative reservoir operating policies were considered in the 1994 Alternatives Report. In particular, the committee could not identify or assess the assumptions made about the initial conditions in Folsom Reservoir and about the operation of Folsom Dam under the various alternatives considered. These concerns are expected to be addressed in upcoming documents, but resolution of these questions should not slow the planning process.

Recommendation

• In the American River planning studies, significant effort should continue to be devoted to development of effective but practical flood control operating policies that make full use of the storage and release capacities of Folsom and other reservoirs in the American River watershed so that the evaluation of alternatives will reflect the level of flood control that the system realistically can achieve and actually should achieve.

Levee Capacity and Long-term Channel Stability in the Lower American River

Currently, there is disagreement among experts about the capabilities of the levees along the lower American River. The committee is concerned about uncertainties related to the proposed alternatives for repairing and enlarging the levees to permit conveyance of "objective releases" from Folsom Reservoir greater than 115,000 cfs. Development of channel and levee stability data should include an understanding of the basic alluvial stratigraphy of the lower American River because the stability of the underlying sediments is critical. Some model and data uncertainty is inevitable, and if that uncertainty about levee adequacy is unacceptable either to the public or decisionmakers, then flood risk reduction alternatives beyond levees—such as building a dam at the Auburn site—may prove unavoidable.

Recommendations

• Before the option of raising and enlarging the levees to permit conveyance of 130,000, 145,000, or 180,000 cfs is included in the flood damage reduction project, the Sacramento District, in concert with SAFCA and other local interests, should ensure that sufficient data and professional consensus concerning the structural stability of specific levees during flow conditions exceeding 115,000 cfs is developed to provide assurance that the levees can contain these higher flows.

• Although it should not slow the decisionmaking process, the Sacramento District, in concert with SAFCA and other local interests, should consider conducting additional work to better understand the long-term geomorphic response of the American River to mining impacts and Folsom Dam because these factors may influence long-term channel stability.

Severity of the American River Flood Risk

Flood risk for the American River is probably greater and more uncertain than indicated by the current estimates. High floods in the latter half of the American River flood series appear to be related to changes in the seasonal distribution of precipitation in the region.

Recommendation

• USACE should assess the magnitude of uncertainty in the American River flood risk and damage estimates by performing a sensitivity analysis involving re-computation of the estimates using just the second half of the American River flood record, from 1950 to the present.

Hydraulic Modeling of the Sacramento-American River System

A better understanding of the complex flow behavior in the neighborhood of Fremont weir and Sacramento weir, Yolo Bypass, and the river junctions between the Feather and Sacramento rivers and the Sacramento and American rivers is needed over the long-term to support water management decision-making in the system.

Recommendations

• The Sacramento District should develop a two-dimensional unsteady flow model of the lower basin to support decisionmaking.
• The Sacramento District should investigate sediment movement and accumulation at weirs, especially during high flows in key areas of the Sacramento and American river system.

Design Considerations for a Dry Dam at Auburn

If constructed without gates, a 425-foot-high dam at the Auburn site would be twice as high as any ungated dam that has been constructed by USACE.

Operational gates are important for dam safety and for providing flexibility in the dam's operation, allowing operators to coordinate with Folsom and other flood control facilities and to minimize environmental impacts in the upper American River canyon by regulating drawdown.

Recommendations

• If a dry dam is built at the Auburn site, it should incorporate operational gates to provide flexibility in the operation of the dam for dam safety considerations, to allow coordination with other facilities, and to minimize environmental impacts.

• Because of the size and possible impacts of a dam at the Auburn site, dam safety studies focused on possible seismic risk should be conducted, reviewed, revised, and acted on as necessary.

• Additional research is needed to better understand the potential impacts of inundation on canyon soils and slope stability, and how those impacts might be mitigated through design or operational considerations.

Hydrologic Monitoring in the Watershed

Soil moisture and snowpack water content affect flood risk in the basin, as does the storage level in reservoirs in the upper American River basin. Measurements of snowpack and streamflow levels are routinely made and used as input to a hydrologic forecasting system operated by the National Weather Service and the California Department of Water Resources that describes the hydrologic status of the basin and provides streamflow forecasts using current hydrologic conditions.

Telemetering of existing streamflow gages and provision of additional telemetered streamflow gages for measuring inflows to Folsom Reservoir are needed to provide reasonable levels of real-time data for reservoir operations and American River basin flood control management. There is also a need for a streamflow gage or improved definition of spillway outflow rating between Folsom and Nimbus dams to ensure accurate control of releases from Folsom.

Recommendations

• Plans to expand the stream-gaging network in place in 1994 should consider installing telemetry in existing gages and providing additional telemetered streamflow gages at the Folsom Lake inlets to provide timely information on the movement of water toward Folsom and downstream toward Sacramento. Plans should also consider providing telemetered capability for accurate, real-time gaging of outflows from Folsom Reservoir. Gages should be strategically located at various elevations within the American River watershed and should be equipped with telemetering equipment to facilitate real-time operations.

• As part of watershed monitoring, the Bureau of Reclamation and USACE should coordinate with the National Weather Service, California Department of Water Resources, U.S. Geological Survey, and others as appropriate in efforts to consider how to use existing data and forecast products and how best to accomplish installation of additional monitoring capacity.

Folsom Operating Guidelines and Training

Lapses or delays in Folsom flood operations can have adverse impacts on system performance, as was the case in the 1986 flood. Operators may be reluctant to rapidly increase or decrease American River discharges as will be required to ensure that high levels of flood risk reduction are achieved by Folsom Reservoir. Steps need to be taken to ensure that lapses in operation do not occur.

Recommendation

• Stricter operating guidelines and operator training, using continuous interactive simulation of different storms, should be implemented to help system operators prepare for and deal with flood events.

Maintaining Efficiency of Flood Control System Operation

Population growth, increasing development, and other changes in the American River watershed create a dynamic flood risk, and it is critical that system operating plans be revisited and revised periodically. For example, as discussed in Chapter 2, it was evident that during the 1986 flood existing operating rules did not adequately consider the existence of 100,000 acre-feet of storage behind the cofferdam upstream of Folsom, even though the dam was designed to breach in a 30-year flood event.

Folsom Reservoir, despite its limitations, is the critical component in the flood control system for Sacramento. Consequently, it is essential that it be operated as efficiently as possible. Potential improvements are being considered in the ongoing development of the Folsom Flood Management Plan. But it is not clear how this ongoing effort considers Folsom reoperation, nor how it considers potential changes in the outlet capacity or storage capacity of the dam, or in the potential construction of upstream storage. It is essential that the operation plan for Folsom evolve in response to changes in the American River flood control system, technological improvements that facilitate reservoir operation, changes in political and economic demands on reservoir storage space, and potential changes in flood regime due to changes in climate and long-term watershed conditions. True efficiency of operations must include all steps from planning through execution, and some long-term mechanism for reviewing operating plans and their implementation is essential.

Recommendations

• The operating plan for Folsom Reservoir should be periodically evaluated and revised as necessary, especially as added experience is gained from extreme events.
• The evaluation and improvement of flood control operating rules for reservoirs such as Folsom should consider contemporary technological capabilities in precipitation and runoff forecasting, remote sensing of rainfall, rainfall-runoff simulation, and real-time monitoring of precipitation, upstream reservoir storage, soil moisture, snowpack, and streamflows. In the future, such capabilities may help operators increase readiness, including temporary encroachment on the conservation and power pool when a major rainfall event is almost certain to occur.

ENVIRONMENTAL ISSUES

Knowledge of Ecosystem Tolerance and Slope Stability

Knowledge of flood tolerances of upland California native plants is limited, and the extent of impacts on Auburn canyon plant communities from a dry dam cannot yet be projected reliably. It can be stated with certainty, however, that even temporary inundation will yield a variety of impacts from submersion, landslides, erosion, and other physical changes that will affect the canyon environment. The nature and extent of impacts cannot be quantified reliably because of the absence of field data on the impacts of periodic inundation on evergreen, physiologically active upland plant communities.

Recommendation

• As long as a dam at the Auburn site remains a proposed or selected alternative, field and laboratory research should continue to better understand the variables affecting the area's plant communities, especially slope stability given fluctuating water levels.

Improving Resource Management

Traditional environmental impact assessments fail to evaluate flood risk management alternatives in an ecosystem context because they use a species-oriented framework. This approach has limited usefulness.

Recommendations

• Environmental impacts should be characterized within the context of the regional significance of the resources involved. Potential scenarios for shifts to different ecosystems or ecotypes and the positive or negative consequences associated with these shifts should be described in environmental impact documents.

• Adaptive management, an approach that includes careful monitoring and opportunities to alter management strategies based on that monitoring, should be incorporated into the American River planning process to allow decisionmakers to proceed with planning while retaining management flexibility. This approach will enhance the mitigation of environmental impacts, even as research to gather new information continues.

Reservoir Operations to Minimize Environmental Impacts

The operating policy for Folsom Dam and a dry Auburn dam, if it were built, would affect the frequency and depth of inundation in the canyon and the rates of drawdown and related impacts. Given the uncertainty of effects in the canyon, a dry dam should be operated jointly with other measures to minimize the frequency of impoundment. A dry dam should not be a first defense against large floods, but rather it should be used to contain peak flows from extreme events.

Recommendation

• Should any alternative that includes a dam at the Auburn site be considered or pursued as a flood risk reduction measure, then the gate design and operating policies should provide options to control the depth and frequency of inundation, allowing operators to reduce plant mortality while keeping drawdown rates low to reduce environmental impacts from landslides.

RISK METHODOLOGY

Risk and Uncertainty Planning Methodologies

The new USACE flood risk and uncertainty analysis procedures are an innovative and timely development that should improve national flood protection planning. The committee sees significant merit in the USACE efforts to better recognize uncertainty in its planning efforts. As time passes, the USACE risk and uncertainty procedures will undoubtedly improve. However, the committee is concerned about the formulation of key steps in the current guidelines and the particular approaches taken to apply the guidelines in the American River planning process.

In particular, the treatment of hydrologic uncertainties, and other potential sources of uncertainty, in the new USACE risk methodology inflates the estimated risk of flooding and the estimates of flood damages that might be avoided by a project. This upward bias is a concern if the methodology is adopted nationwide because it could distort the economic evaluation of projects. (The committee did not have the resources to determine or evaluate the actual distortion for the American River study.)

Recommendations

• USACE needs to develop a consistent scientific methodology and an effective vocabulary for description of residual flood risks and uncertainties to technical and public audiences.

• To avoid the problem of bias described above, and to simplify the analysis so that it can be more easily understood, the primary descriptions of the expected economic damages and the probability of flooding should be based on traditional estimates of the flood-flow frequency relations and other factors that contribute to flood risk and damage, such as those given in Bulletin 17B, without the expected probability or other uncertainty adjustments.

• Best estimates of expected annual damages and the risk of flooding should be supplemented by descriptions of the uncertainty or the possible errors in such performance criteria due to hydrologic, hydraulic, and economic uncertainties. Uncertainty and its impact can be described by a standard error or percentiles of the distribution of system performance criteria due to uncertainty, or the probability that the net economic benefits might turn out to be less than zero. This would be consistent with the requirement in the USACE risk and uncertainty procedures that "the estimate of NED (National Economic Development) benefits will be reported both as a single expected value and on a probabilistic basis . . . for each planning alternative." It is the committee's understanding the American River study will not address economic uncertainties.

Estimates of expected damages and economic benefits associated with different projects, and the probability of flooding at different locations, are likely to be the primary criteria describing flood risk and economic impacts. It will often be useful to calculate other indices of system performance and the reliability of different components of the river channel and levee system. The committee questions in general the value of the system "reliability" index adopted by the USACE and employed in the American River study. It seems to be an awkward combination of traditional and new concepts that would easily be misunderstood.

Ecological Risk Assessment

Ecological risk assessment is not yet sufficiently developed to provide much useful guidance for evaluation of flood control alternatives in the American River planning effort at this time. However, it does provide a new approach to help us think about environmental impacts and select questions for investigation; it will be increasingly important in helping planners broaden the context examined in future planning activities.

Recommendation

• USACE should follow the rapidly evolving potential of ecological risk assessment and adopt this approach as it develops to improve the decisionmaking process.

FLOOD RISK MANAGEMENT BEHIND LEVEES

Federal Participation in Flood Damage Reduction Projects for the American River

Development within the levees of the Natomas Basin faces chronic flood risk. Improvements in the existing flood protection system including the reoperation of Folsom Dam and levee expansion are in progress or foreseeable. Other measures that might contribute to flood hazard reduction, such as construction of a dry flood storage dam at Auburn, are hypothetical and speculative at this writing. Environmental, fiscal, and political factors are likely to continue to delay or even eliminate that option from consideration. The future level of reliable flood protection therefore is difficult if not impossible to assess in light of both hydrologic and sociopolitical uncertainty.

Those who propose to permit new development in the Natomas basin should not assume that federal flood control projects on the American River will eliminate flood risk or that the only flood risk is from the American River. Additional hazards are presented by the Sacramento River and by limited internal drainage capabilities. In addition, flood risk will continue for the already developed parts of the city. The committee does not sanction the development of Natomas, but in acknowledgment of the development pressures and in recognition of extensive existing development, the committee recommends that future federal participation in flood damage reduction projects for the American River be conditioned upon the following:

Recommendations

• Congress should explicitly determine whether flood control projects on the American River warrant federal involvement based on the presence of widespread national benefits from flood protection or on a limited ability of the community to provide its own flood protection.

• If a federal interest in flood protection works in the American River basin is established, project construction should be delayed until SAFCA, working with FEMA and private insurers, has a program to require new development at Natomas and in the city to purchase flood insurance at actuarially sound rates for the residual flood risk appropriate to the alternative selected.

•   SAFCA should implement a flood hazard mitigation plan that includes flood risk communication, flood warning systems, evacuation plans to reduce loss of life, highway and other infrastructure designs to facilitate evacuation, and floodproofing and elevation requirements wherever cost effective. The public should be informed of the inherent flood risks pertaining to the Natomas Basin despite the levee system.

Reforming National Flood Risk Management Policy

A recent report, *Sharing the Challenge* (IFMRC, 1994), suggested that the standard project flood (defined as the "most significant flood event" expected to occur) be adopted as the basis for planning a comprehensive flood risk management program. To obtain maximum benefits, a comprehensive flood risk management strategy should consider nonstructural flood damage reduction measures together with structural measures, especially as applicable in currently undeveloped areas such as the Natomas Basin. These measures should include appropriate floodplain zoning, floodproofing, education, and, when feasible, relocation. A comprehensive program would ensure that people who locate in hazardous areas bear, to the extent practicable, the costs of that location decision by paying a substantial share for flood water control works, by accepting restrictions on development in flood prone areas (foregone development value), and by paying adequate insurance premiums against the residual flood risk after structural and nonstructural measures have been implemented.

Recommendations

•   USACE, FEMA, and other relevant federal agencies should adopt an agreement governing federal participation in structural and nonstructural flood risk management efforts to require that benefiting local communities have a program requiring new development to purchase flood insurance at actuarially sound rates for residual flood risk. Existing development should also purchase residual risk insurance. The federal government, possibly working with private insurers, should develop provisions for sharing the cost of flood insurance premiums and flood damage reduction measures with communities and individuals who implement structural and nonstructural flood damage reduction measures.

•   Congress should reform the cost-sharing requirements in the 1986 Water Resources Development Act to increase the nonfederal cost share above currently authorized levels. Exemptions to the cost share requirement could be made if it is demonstrated that flood control benefits are widespread, or that the benefiting communities have limited ability to pay for otherwise justified flood protection, or that communities have committed to flood mitigation programs assigned a high rating in the National Flood Insurance Program's community rating system, thereby exceeding minimum federal criteria for floodplain management.

• Congress should define all cost-sharing requirements as a percentage of total project costs. Under special conditions when the local sponsor can demonstrate both that noncash contributions (e.g., lands, easements, and rights of way) are necessary for the project and that they will be undertaken, such contributions should be allowed to offset nonfederal cost responsibilities.

WATER RESOURCES PLANNING AND DECISIONMAKING

Risk Communication for the American River

The Sacramento District continues to use the term "level of protection" in its presentations and publications and has not developed good explanations of its new risk and uncertainty procedures and of the results from their application in the American River. However, sound risk communication is of central importance to effective risk management.

Recommendations

• The Sacramento District should cease using "level of protection" and act with SAFCA and other local leaders to build and publicize realistic scenarios to describe the consequences of a levee being overtopped or breached, with the goal of educating the public about the risk and increasing preparedness. Description of the vulnerability of the Sacramento and Natomas areas to storm events that overtop or breach the levee system should clearly address the extreme depth of flooding possible, the transportation difficulties that will be faced, and the problems recovering from flooding in a closed basin.

• As the change-over in terminology occurs, and "level of protection" continues to be used, the Sacramento District and SAFCA should interpret the concept of 100-year "level of protection" in ways that more clearly articulate the risks in terms with meaning to the public. For example, using the old terminology, a 100-year "level of protection" includes a 40 percent chance of at least one catastrophic flood event in the next 50 years. Thus the probability of at least one catastrophic flood within the lifetime of most residents is roughly equal to the probability of flipping a coin and getting heads. Similarly, the Sacramento District and SAFCA should interpret the flood risk on the American River with a 200-year "level of protection" as a 22 percent chance over the next 50 years that flood waters will overtop the levees and inundate the Sacramento area.

Risk Communication in Federal Programs

USACE, as an agency, has great national influence in flood management planning. Its new risk and uncertainty assessment procedures should contribute to understanding the likelihood and consequences of flood events under different risk management alternatives. The traditional term "level of protection" may mislead the public and is not consistent with the analytical outcomes expected from the new USACE procedures.

Recommendations

- USACE should select a technically sound risk communication vocabulary and approach to communicating flood risk likelihood and consequences and use it consistently in all reports and presentations.
- USACE should work with FEMA, as well as other agencies, the states, and private insurers, to develop a standardized vocabulary that adequately conveys risk and vulnerability from flooding.

Improving USACE Risk and Uncertainty Assessment and Risk Communication

Several of the findings and recommendations listed above address the committee's concerns with the USACE approach to risk and uncertainty analysis and risk communication, as these were described to the committee. The risk and uncertainty procedures are clearly new and still under development. In fact, the American River is one of the first applications, and almost surely the most complex yet attempted. Some committee members questioned whether the methodology was sufficiently well developed to be adopted as the basis of the evaluation of such an important and controversial project. On the other hand, there is a tradition in USACE and among engineers to work out the details of analysis methodologies as they are being implemented.

Recommendation

- USACE should convene an intra-agency workshop that would include invited outside experts familiar with risk communication issues and risk and uncertainty procedures related to water resources projects. The purpose of the workshop would be to review the concerns expressed by this committee and others pertaining to the new USACE approach to risk and uncertainty analysis, as well as the USACE approach to the communication of those results to technical and general audiences. The workshop should develop guidelines that have broad support among risk analysis experts within and outside USACE.

Planning for the American River

Failure by SAFCA and the Sacramento District to incorporate a wide range of purposes and institutional adjustments into plan formulation and to open the American River planning process to multiple interests was a barrier to reaching agreement on a flood risk management alternative.

Recommendations

• Flood risk management decisions for the American River are influenced by the larger context of California water management, including several recent environmental rulings, and can be addressed properly only within that context. As a result, the state of California should not expect the flood control controversy on the American River to be resolved solely under the Sacramento District's leadership. The state needs to increase its participation in, and resources dedicated to, basin-wide water resources planning. State leadership is critical to build a consensus on technical and institutional strategies to manage competing water demands.

• If no agreement is reached on an acceptable approach to flood risk management in the near future, the Sacramento District and SAFCA should expand the consensus-building efforts of the Lower American River Task Force. This expanded effort should address the full array of purposes that were originally part of the study authorization, including water supply and allied purposes. It should also work to identify institutional agreements that can be employed to address these purposes. In this effort, the Sacramento District and SAFCA might request the leadership and assistance of the state of California's Resources Agency.

The American River as Part of a Larger California Water System

The proposed shift in release patterns from Folsom Reservoir related to reoperation to reflect upstream storage levels may affect seasonal releases into the Sacramento River and the delta system. Changes in releases have implications that go beyond the American River basin. Likewise, the Central Valley Project Improvement Act (P.L. 102-575, Title 34) and the federal-state water quality standards announced in December 1994 for the San Francisco/Sacramento-San Joaquin delta estuary may affect Folsom operations and water release requirements.

Recommendation

• Development of flood control operating guidelines for the American River needs to recognize the wider impacts of reoperation of storage facilities in the

American River basin, including revised downstream water quality and habitat requirements.

Reforms in Federal Planning

Plan formulation demands the creation of the widest possible range of engineering and institutional alternatives so that agreements can be reached among multiple decisionmakers.

Recommendation

- USACE should issue guidance to its districts stressing the requirement to maintain a broad view of water resource planning purposes and address those multiple purposes throughout the planning process, including both the development and evaluation of institutional as well as engineering measures. Especially when contentious disagreements are involved, USACE should advise its districts to facilitate but not dominate local decisionmaking.

CONCLUSION

This committee was charged to evaluate the scientific and engineering knowledge on which the selection of a flood hazard reduction strategy for the lower American River will ultimately be based, and to provide insights where possible on public policies concerning flood hazard management in the Unites States. To these ends, the committee has presented more than 20 findings and associated recommendations. These are offered in a spirit of constructive criticism and to encourage the continued progress in reducing the American River basin's flood risk and in the evolution of the nation's understanding of flood risk management. The committee reiterates its concern that nothing stated in this report should be used as an excuse for delaying action in the American River basin. It is time to select and implement appropriate flood risk reduction strategies.

References

Aguado, E., D. Cayan, L. Riddle, and M. Roos. 1992. Climatic fluctuations and the timing of West Coast streamflow. Journal of Climate 5(12):1468-1483.

Arnell, N. 1989. Expected annual damages and uncertainties in flood frequency estimation. Journal of Water Resources Planning and Management 115(1):94-107.

Bailey, R.G., and R.M. Rice. 1969. Soil slippage: An indicator of slope instability on chaparral watersheds of southern California. Professional Geography 21:172-177.

Barbour, M., B. Paulik, F. Drysdale, and S. Lindstrom. 1993. California's Changing Landscapes: Diversity and Conservation of California Vegetation. Sacramento: California Native Plant Society.

Beard, L.R. 1960. Probability estimates based on small normal-distribution samples. Journal of Geophysical Research 65(7):2143-2148.

Beard, L.R. 1962. Statistical Methods in Hydrology, January 1962. Sacramento: USACE.

Beard, L.R. 1978. Impact of hydrologic uncertainties on flood insurance. Journal of the Hydraulics Division of the ASCE 104(HY11):1473-1483.

Beard, L.R. 1990. Discussion of expected annual damages and uncertainties in flood frequency estimation. Journal of Water Resources Planning and Management 116(6):847-850.

Benjamin, J.R., and C.A. Cornell. 1970. Probability, Statistics and Decision for Civil Engineers. New York: McGraw-Hill.

Bernier, J.M. 1987. Elements of Bayesian analysis of uncertainty in hydrological reliability and risk models. Pp. 405-422 Engineering Reliability and Risk in Water Resources (NATO Advanced Study Institute, Tucson, Ariz., May 1985), L. Duckstein and E. Plate (eds.). Dordrecht, the Netherlands: M. Nijhoff.

Bischofberger, T.E. 1975. Early flood control in the California Central Valley. Journal of the West 14:85-94.

Booy, C., and L.M. Lye. 1989. A new look at flood risk determination. Water Resources Bulletin 25(5):933-944.

Bowles, D.S. 1990. Risk assessment in dam safety decisionmaking. Pp. 254-283. in Y.Y. Haimes and E.Z. Stakhiv (eds.). Risk-based Decision Making in Water Resources. New York: American Society of Civil Engineers.

Brooks, S.M., K.S. Richards, and M.G. Anderson. 1993. Approaches to the study of hillslope development due to mass movement. Progress in Physical Geography 17(1):32-49.

Brunsden, D. 1979. Mass movements. Pp. 187-212 in Process in Geomorphology, C. Embleton and J. Thornes (eds.). New York: John Wiley & Sons.

Bureau of the Census. 1991. Statistical Abstract of the United States. Washington, D.C.: U.S. Government Printing Office.

Burges, S.J. 1979. Analysis of uncertainty in flood plain mapping. Water Resources Bulletin 15(1):227-243.

Burkham, D.E. 1978. Accuracy of flood mapping. Journal of Research, U.S. Geological Survey 6(4):515-527.

California Debris Commission. 1907. Map of American River, California, from its mouth in the Sacramento River to the South Fork. U.S. Army Corps of Engineers; Tower, M.L., Jr. Engineer, 7 sheets plus cross sections 1:9600.

California Department of Fish and Game (CDFG), Inland Fisheries Division. 1993. Restoring Central Valley Streams: A Plan for Action. Sacramento: CDFG.Cedergren, H.R. 1989. Seepage, Drainage, and Flow Nets. New York: John Wiley & Sons.

California Department of Water Resources (CDWR). 1986. California High Water, Bulletin 69-86. Sacramento: State of California.

California Department of Water Resources (CDWR). 1988. Dams Within Jurisdiction of the State of California. Bulletin 17-88. 121 pp.

California Department of Water Resources (CDWR). 1991. Evaluation of soils and soil stability for the proposed flood control dam at Auburn. In American River Watershed Investigation, California: Feasibility Report Appendix M, Chapter 8. Sacramento: USACE.

California Resources Agency. 1993. The Yolo Basin wetlands. Golden State Floodlight 8(1).

California Resources Agency, and California Department of Water Resources. 1993. "The Yolo Basin Wetlands," Golden State Floodlight, Vol. 8 No. 1 Winter/Spring 1992-93.

California State Lands Commission. 1994. California's Rivers: A Public Trust Report. Sacramento: California State Lands Commission.

California State Water Resources Control Board (CSWRCB). 1994. Comments on EPA's Proposed Criteria for the Bay-Delta Estuary. March 11, 1994. Sacramento: CSWRCB.

Campbell, R.H. 1975. Soil slips, debris flows, and rainstorms in the Santa Monica Mountains and vicinity, southern California. U.S. Geological Survey Professional Paper 851. 51 pp.

Carson, M.A., and D.J. Petley. 1970. The existence of threshold slopes in the denudation of the landscape. Transactions of the Institute of British Geographers 49:71-95.

Chandler, R.J. 1986. Processes leading to landslides in clay slopes: A review. Pp. 343-360 in Hillslope Processes, A.D. Abrahams (ed.). Boston: Allen & Unwin.

Chasse, E., and G.A.J. Platenkamp. 1994. American River Flood Control Project Special Evaluation—Vegetation Inundation-Mortality Study of the Proposed Auburn Flood Control Facility. Montgomery Watson Americas, Inc., and Jones and Stokes Associates. Prepared for the Sacramento District, USACE, Sacramento, Calif.

Chow, V.T., D.R. Maidment, and L.W. Mays. 1988. Applied Hydrology, New York: McGraw-Hill.

Chowdhury, J.U., and J.R. Stedinger. 1991. Confidence interval for design floods with estimated skew coefficient. Journal of Hydraulic Engineering 117(7):811-831.

Chowdhury, R.N. 1992. Simulation of risk of progressive slope failure. Canadian Geotechnical Journal 29:94-102.

City of Sacramento. 1990. Land Use Planning Policy Within the 100 Year Flood Plain in the City and County of Sacramento. Planning and Development Department. M89-054.

City of Sacramento. 1993a. Land Use Planning Policy Within the 100 Year Floodplain. October 12. Report #3. Planning and Development Department. Mimeograph.

City of Sacramento. 1993b. Draft North Natomas Community Plan. May 1986, amended 1993.

City of Sacramento. 1994. Status of Comprehensive Flood Management Plan. Department of Utilities. Memo. April 12.

Cohrssen, J.J., and V.T. Covello. 1989. Risk Analysis: A Guide to Principles and Methods for Analyzing Health and Environmental Risks. Washington, D.C.: U.S. Council on Environmental Quality, Executive Office of the President.

Congressional Record. 1992. October 8. S17910.

Cooke, R.U., and J.C. Doornkamp. 1990. Geomorphology in Environmental Management: A New Introduction. Second Ed. Oxford: Clarendon Press.

Countryman, J.D. 1993. Folsom Dam Flood Management Report, Murray, Burns and Kienlen Consulting Civil Engineers, DRAFT Report to Sacramento Area Flood Control Agency. July.

Council on Environmental Quality. 1973. Federal Guidelines on Alternatives to the Proposed Action, Federal Register, May 2, 1973.

Cunnane, C. 1988. Methods and merits of regional flood frequency analysis. Journal of Hydrology 100:269-290.

Davis, D.R., C.C. Kisiel, and L. Duckstein. 1972. Bayesian decision theory applied to design in hydrology. Water Resources Research 8(1):33-41.

Davis, D.W. 1991. A risk and uncertainty based concept for sizing levee projects (reprinted in course notes for Risk-based Analysis for Flood Reduction Projects, USACE Hydrologic Engineering Center, Davis, Calif., May 2-6, 1994). In Proceedings of a Hydrology and Hydraulics Workshop on Riverine Levee Freeboard, SP-24. Monticello, Minn.

DeGraff, J.V. 1994. The geomorphology of some debris flows in the southern Sierra Nevada, California. Pp. 231-252 in Geomorphology and Natural Hazards, M. Morisawa (ed.). New York: Elsevier.

Dickert, T.G., and K.R. Domeny (eds.). 1974. Environmental Impact Assessment: Guidelines and Commentary. Berkeley: The Regents of the University of California.

Dillinger, W.C. (ed.) 1991. A History of the Lower American River. American River Natural History Association. 165 pp.

Dingman, S.L., and R.H. Platt. 1977. Floodplain zoning: Implications of hydrologic and legal uncertainty. Water Resources Research 13(3):519-523.

Doran, D.G., and J.L. Irish. 1980. On the nature and extent of bias in flood damage estimation. Paper presented at the Hydrology and Water Resources Symposium, Adelaide, Australia, November 4-6.

Dotson, H.W., D.W. Davis, M.T. Tseng, and E.E. Eiker. 1994. Case Study: Risk-based Analysis of Flood Reduction Measures. Budapest: NATO Advanced Study Institute for Floodplain Management.

Duckstein, L., and J. Bernier, 1986. A System Framework for Engineering Risk Analysis. Pp. 90-110 in Risk-based Decision Making in Water Resources, Y.Y. Haimes and E.Z. Stakhiv (eds.). (Proceedings of an Engineering Foundation Conference, Santa Barbara, Calif., November 3-5, 1985.) New York: American Society of Civil Engineers.

Duckstein, L., I. Bogárdi, F. Szidarovszky, and D.R. Davis. 1975. Sample uncertainty in flood levee design: Bayesian versus non-Bayesian methods. Water Resources Bulletin 11(3):425-435.

Duckstein, L., E. Plate, and M. Benedini. 1987. Water engineering reliability and risk: A system framework. In Engineering Reliability and Risk in Water Resources (NATO Advanced Study Institute, Tucson, Ariz., May 1985), L. Duckstein, and E. Plate (eds.). Dordrecht, the Netherlands: M. Nijhoff.

Dyhouse, G.R. 1985. Stage-frequency analysis at a major river junction. Journal of Hydraulic Engineering of the ASCE 111(4):565-583.

EIP Associates. 1992. Sacramento Area Flood Control Agency Swainson's Hawk and Giant Garter Snake Draft Habitat Conservation Plan. Sacramento, Calif. February.

Earle, C. 1993. Asynchronous droughts in California streamflow as reconstructed from tree rings. Quaternary Research 39:290-299.

Ellen, S.D., and R.W. Fleming. 1987. Mobilization of debris flows from soil slips, San Francisco Bay region, Calif. Pp. 31-40 in Debris Flows/Avalanches: Process, Recognition, and Mitigation, J.E. Costa and G.F. Wieczorek (eds.). Boulder, Colo.: Geological Society of America.

Environmental Defense Fund. 1990. Critique of U.S. Army Corps of Engineers Sacramento District Office Analyses of Flood Control for the American River. Oakland, Calif.

Ervine, D.A., B.B. Willetts, R.H.J. Sellin, and M. Lorena. 1993. Factors affecting conveyance in meandering compound flows. Journal of Hydraulic Engineering 119:1383-1399.

Estes, G. 1993. Memo. Figures 4 and 5 based on FEMA and California Department of Water Resources data. August 16.

Evans, S.G. 1986. Landslide damming in the Cordillera of western Canada. Pp. 111-130 in Landslide Dams: Processes, Risk, and Mitigation, R.L. Schuster (ed.). American Society Civil Engineers, Geotechnical Special Publication 3.

Farquhar, F.P. 1932. The topographical reports of Lieutenant George H. Derby. Quarterly of the California Historical Society 11(2):117.

Federal Emergency Management Agency (FEMA). 1994. A Unified National Program for Floodplain Management. FEMA. Washington, D.C. Part of a series originating in 1966.

Fisher, F.W. 1994. Past and present status of Central Valley chinook salmon. Conservation Biology 8:870-873.

Foster, H.A. 1924. Theoretical frequency curves and their application to engineering problems. American Society of Civil Engineers Trans. 87:142-173.

Frederick, S.W., and R.M. Peterman. 1995. Choosing fisheries harvest policies: When does uncertainty matter? Canadian Journal of Fisheries and Aquatic Sciences 52(2): in press.

Freeze, R.A. 1987. Modelling the interrelationships between geology, hydrogeology, and climate. Pp. 381-403 in Slope Stability. M.G. Anderson and K.S. Richards (eds.). New York: John Wiley and Sons.

Fugro-McClelland (West), Inc., and A.T. Leiser. 1991. Inundation Impact Analysis. In American River Watershed Investigation, Calif. Feasibility Report, Appendix Q. Sacramento: USACE.

Gerstung, E.R. 1971. A Report to the California State Water Resources Control Board on the Fish and Wildlife Resources of the American River to be Affected by the Auburn Dam and Reservoir and the Folsom South Canal and Measures Proposed to Maintain These Resources. Sacramento: California Department of Fish and Game, Region 2.

Gilbert, G.K. 1917. Hydraulic Mining debris in the Sierra Nevada. U.S. Geological Survey Professional Paper 105. 154 p.

Goicoechea, A., C. Whipple, G.P. Johnson, M. Collins, and M.M. Laws. 1987. Risky waters. Civil Engineering April, 68-71.

Gould, B.W. 1973. Discussion of bias in computed flood risk by C.H. Hardison and M.E. Jennings. Journal of the Hydraulics Division of the ASCE 99(HY1):415-427, 1973.

Graf, W.L. 1977. The rate law in fluvial geomorphology. American Journal of Science 277:178-191.

Haimes, Y.Y., and E.Z. Stakhiv (eds.). 1986. Risk-based Decision Making in Water Resources. New York: American Society of Civil Engineers.

Haimes, Y.Y., and E.Z. Stakhiv (eds.). 1989. Risk-based Analysis and Management of Natural and Man-made Hazards. New York: American Society of Civil Engineers.

Haimes, Y.Y., and E.Z. Stakhiv (eds.). 1990. Risk-based Decision Making in Water Resources. New York: American Society of Civil Engineers.

Haimes, Y.Y., D.A. Moser, and E.Z. Stakhiv (eds.). 1992. Risk-based Decision Making in Water Resources. New York: American Society of Civil Engineers.

Hardison, C.H., and M.E. Jennings. 1972. Bias in computed flood risk. Journal of the Hydraulic Division of the ASCE 98(HY3):415-427.

Hardison, C.H., and M.E. Jennings. 1973. Bias in computed flood risk-closure. Journal of the Hydraulic Division of the ASCE 99(HY7):1157-1158.

Harris, T. 1986. A flood tide of criticism for federal water officials. Sacramento Bee. February 24.

Hart, J., R. Meredith, and J. Keeley. 1994. Tolerance of Plants to Deepwater Flooding. Draft Report to the Sacramento Area Flood Control Agency.

Heyman, I.M. 1974. NEPA/CEQA: Legal aspects. In Environmental Impact Assessment: Guidelines and Commentary, T.G. Dickert and K.R. Domeny (eds.). Berkeley: The Regents of the University of California.

Hicks, L., and A. D. Blechman. 1994. Sacramento Bee, December 2, 1994. B1, B6.

Hilborn, R., and C.J. Walters. 1992. Quantitative Fisheries Stock Assessment: Choice, Dynamics and Uncertainty. New York: Chapman and Hall. 570 pp.

Hirschboeck, K.K. 1988. Flood hydroclimatology. In Flood Geomorphology. Baker, V.R., R.C. Kochel, and P.C. Patton (eds.). New York: John Wiley. Pp. 27-49.

Holling, C.S. (ed.) 1978. Adaptive Environmental Assessment and Management. Chichester: John Wiley.

Interagency Advisory Committee on Water Data (IACWD). 1982. Guidelines for determining flood flow frequency, Bulletin 17B of the Hydrology Subcommittee. Reston, Va.: USGS.

ISO Technical Advisory Group on Meteorology (TAG 4). 1993. Guide to the Expression of Uncertainty in Measurement. Geneva: International Organization for Standardization. 101 pp.

Interagency Floodplain Management Review Committee. 1989. Action Agenda for Managing the Nation's Floodplains. App. F of Floodplain Management in the United States, Vol. 2: Full Report. Washington: Federal Interagency Floodplain Management Task Force.

Interagency Floodplain Management Review Committee. 1994. Sharing the Challenge: Floodplain Management into the 21st Century. Washington, D.C.: U.S. Government Printing Office.

Interagency Floodplain Management Task Force. 1992a. Floodplain Management in the United States: An Assessment Report. Vol. 1. Summary Report. Washington D.C.: U.S. Government Printing Office.

Interagency Floodplain Management Task Force. 1992b. Floodplain Management in the United States. Vol. 2. Full Report. Washington, D.C.: U.S. Government Printing Office.

Iverson, R.M., and R.G. LaHusen. 1989. Dynamic pore-pressure fluctuations in rapidly shearing granular materials. Science 246:796-798.

Jackson Daily News. August 27, 1964. Devastating Floods Recalled in Flowood. B-10.

Jaguette, D. L. 1978. Efficient Water Use in California: Conjunctive management of ground and surface reservoirs. R-2389-CSA/RF. Santa Monica, Calif.: Rand.

James, L.A. 1991a. Incision and morphologic evolution of an alluvial channel recovering from hydraulic mining sediment. Geological Society of America Bulletin 103:723-736.

James, L.A. 1991b. Quartz concentration as an index of sediment mixing: hydraulic mine-tailings in the Sierra Nevada, Calif. Geomorphology 4:125-144.

James, L.A. 1994. Channel changes wrought by gold mining: Northern Sierra Nevada, California. Proceedings of the American Water Resources Association; Symposium, Jackson Hole, WY. In Marston, R.A. and Hasfurther, V. (ed.), Effects of Human-Induced Changes on Hydrologic Systems.

James, L.B., and G. Kiersch. 1991. Failures of engineering works. Pp. 480-516 in The Heritage of Engineering Geology: The First Hundred Years. Centennial Special Vol. 3. Geological Society of America.

Jennings, J. 1991. Comments on the American River Watershed Investigation Draft Feasibility Report, Environmental Impact Statement and E.I.R. Sacramento, Calif.: Planning and Conservation League. June 14.

Jennings, J. 1992. Comments on Authorizing Legislation. Planning and Conservation League, Sacramento. March 12. Memorandum to SAFCA Board.

Keeley, J.E. 1992. Comments on Fugro-McClelland and Leiser Report of Dec. 1991. Submitted to Sacramento Area Flood Control Agency. Los Angeles, Calif.: Occidental College. January.

Kelley, R. 1989. Battling the Inland Sea: American Political Culture, Public Policy, and the Sacramento Valley, 1850—1986. Berkeley: University of California Press. 395 pp.

Knighton, A.D. 1974. Variation in Width-Discharge relation and some implications for hydraulic geometry. Geological Society of America Bulletin 85:1069-76.

Knox, J.C. 1983. Responses of river systems to Holocene climates. In Late-Quaternary Environments of the United States. Wright, H.E. Jr., (ed.). Minneapolis: University of Minnesota Press. Vol. 2, pp. 26-41.

Knudsen, M.D. 1991. Potential vegetation and wildlife impacts of the proposed Auburn dry dam: A literature review and discussion of similar peak flow facilities. In American River Watershed Investigation Feasibility Report. Vol. 6, App. S. U.S. Fish and Wildlife Service. Sacramento, Calif.: USACE.

Kuczera, G. 1982. Combining site-specific and regional information, an empirical Bayes approach. Water Resources Research 18(2):306-314.

Kuczera, G. 1988. On the validity of first-order prediction limits for conceptual hydrologic models. Journal of Hydrology 103:229-247.

Lackey, R.T. 1994. Ecological risk assessment. Fisheries 19(9):14-18.

Lettenmaier, D.P. and T.Y. Gan. 1990. Hydrologic sensitivities of the Sacramento-San Joaquin River basin, California, to global warming. Water Resources Research 26(1):69-86.

Leopold, L.B. 1969. Quantitative comparison of some aesthetic factors among rivers. U.S. Geological Survey Circular 620. Washington, D.C.

Leopold, L.B. 1974. The Use of Data in Environmental Impact Assessment. In Environmental Impact Assessment: Guidelines and Commentary. T.G. Dickert and K.R. Domeny (eds.). Berkeley: The Regents of the University of California.

Leopold, L.B., F.E. Clarke, B.B. Hanshaw, and J.R. Balsley. 1971. A procedure for evaluating environmental impact. U.S. Geological Survey Circular 645. Washington, D.C.

Lord, W. 1979. Conflict in federal water resource planning. Water Resources Bulletin 15:1226-1235.

Lower American River Task Force. 1994. Proceedings of Phase One: The Lower American River Task Force, Principals and Guidelines for Flood Control Planning Along the Lower American River. Sacramento Area Flood Control Agency, Sacramento, California.

Ludwig, D., R. Hilborn, and C. Walters. 1993. Uncertainty, resource exploitation, and conservation: Lessons from history. Science 260:17, 36.

Manson, M. 1882. Report of Mr. Marsden Manson, Assistant Engineer. Mendell, Col. G. H., Report upon a Project to Protect the Navigable Waters of Calif. from the Effects of Hydraulic Mining. House Doc. 98, 47th Congress, 1st Session:78-101.

Mays, L.W., and Y.K. Tung. 1992. Hydrosystems Engineering and Management. New York: McGraw-Hill.

McCann, M.W., Jr., J.B. Franzini, E. Kavazanjian, and H.C. Shah. 1985. Preliminary Safety Evaluation of Existing Dams, Vols. I and II, Report 69-70. Stanford, Calif.: John Blume Earthquake Engineering Center, Stanford University.

McCuen, R.H. 1979. Journal of the Water Resources Planning and Management Division of the ASCE 105(WR2):269-277 (with Closure 107(WR2):582, 1981).

Mendell, Col. G.H. 1881. Protection of the navigable waters of California from injury from the debris of mines. House Document 76, 46th Congress, 3rd Session.

Meredith, R.W., A.T. Leiser, and Fugro-West Inc. 1994. Short- and Long-Term Impacts of Periodic Flooding on Chaparral and Oak Woodland Species Along the Upper Sacramento River. Draft report to the Sacramento Area Flood Control Agency. June.

Moran, P.A.P. 1957. The statistical treatment of floods. Transactions of the American Geophysical Union 38(4):519-523.

Morgan, M.G., and M. Henrion. 1990. Uncertainty: A Guide to Dealing with Uncertainty in Quantitative Risk and Policy Analysis, Cambridge University Press.

Moser, D.A. 1994. Risk analysis framework for evaluation of hydrologic/hydraulics and economics in flood damage reduction studies. Course notes for Risk-based Analysis for Flood Reduction Projects, USACE Hydrologic Engineering Center, Davis, Calif., 2-6 May.

Munz, P., and D.D. Keck. 1973. California Flora and Supplement. Berkeley: University of California Press.

National Research Council (NRC). 1983a. Risk Assessment in the Federal Government: Managing the Process. Washington, D.C.: National Academy Press.

National Research Council (NRC). 1983b. Safety of Existing Dams: Evaluation and Improvement. Washington, D.C.: National Academy Press.

National Research Council (NRC). 1985. Safety of Dams: Flood and Earthquake Criteria. Washington, D.C.: National Academy Press.

National Research Council (NRC). 1989. Improving Risk Communication. Washington, D.C.: National Academy Press.

National Research Council (NRC). 1993. Issues in Risk Assessment. Washington, D.C.: National Academy Press.

National Research Council (NRC). 1994. Science and Judgement in Risk Assessment. Washington, D.C.: National Academy Press.

Neill, C.R. (ed.). 1990. Stability of Flood Control Channels. Draft document prepared for U.S. Army Corps of Engineers, Waterways Experiment Station and Committee on Channel Stabilization. Washington, D.C: USACE.

Olmsted, F.H., and G.H. Davis. 1961. Geologic features and ground-water storage capacity of the Sacramento Valley, California. U.S. Geological Survey Water Supply Paper 1497.

Paul, R.W. 1947. California Gold: The Beginning of Mining in the Far West. Lincoln: University of Nebraska Press. 380 pp.

Peterman, R.M. 1990. Statistical power analysis can improve fisheries research and management. Canadian Journal of Fisheries and Aquatic Sciences 47:2-15.

Platt, R.H. 1982. The Jackson Flood of 1979: A public policy disaster. Journal of the American Planning Association 48(2):219-231.

Platt, R.H. 1994. Law: Parsing Dolan. Environment 36(8):4-5.

Plough, A., and S. Krimsky. 1987. The emergence of risk communication studies: Social and political context. Science, Technology and Human Values 12(3-4):4-10.

Porterfield, G., R.D. Busch, and A.O. Waananen. 1978. Sediment transport in the Feather River, Lake Oroville to Yuba City, Calif. U.S. Geological Survey Water-Resources Investigation 78-20:73.

Potter, K.W. 1987. Research on flood frequency analysis: 1983-86. Reviews of Geophysics 25(2):113-118.

Potter, K.W., and D.P. Lettenmaier. 1990. A comparison of regional flood frequency estimation methods using a resampling method. Water Resources Research 26(3):415-424.

Pupacko, A. 1993. Variations in northern Sierra Nevada streamflow: implications of climate change, Water Resources Bulletin 29(2):283-290.

Redmond, K.T. and R.W. Kock. 1991. Surface climate and streamflow variability in the Western United States and their relation to large-scale circulation indices. Water Resources Research 27(9):2381-2399.

Remy, M.H., J.G. Moose, and T.A. Thomas. 1994. Guide to the California Environmental Quality Act. 8th Edition. Berkeley, California: Solano Press.

Resource Consultants and Engineers, Inc. 1993. American and Sacramento River, California Project; Geomorphic, Sediment Engineering and Channel Stability Analyses. Draft Report for U.S. Army Corps of Engineers contract DACW05-93-C-0045.

Rezac, F. 1993. Painted Rock Dam Controls Arizona Flood. In USCOLD Newsletter, United States Committee on Large Dams, Issue No. 101, July.

Rice, R.M. 1982. Sedimentation in the chaparral: How do you handle unusual events? Pp. 39-49 in Sediment Budgets and Routing in Forested Drainage Basins, F.J. Swanson et al. (eds.). USDA Forest Service, Pacific Northwest Forest and Range Experiment Station, General Technical Report PNW-141. Portland, Ore.:

Riggs, H.C. 1966. Chapter 3—Frequency curves. In U.S. Geological Survey Surface Water Techniques Series, Book 2. Washington, D.C.: U.S. Geological Survey.

Riley, A.L. 1974. Water Quality and Environmental Study for Health and Land Use Planning, Johnson County, Iowa.

Riley, A.L. 1989. Overcoming federal water policies, the Wildcat-San Pablo Creeks case. Environment 31(10).

Roos, M. 1987. Possible changes in California snowmelt runoff patterns. Pp. 22-31 in Proceedings, Fourth Annual PACLIM Workshop, Pacific Grove CA.

Roos, M. 1994. Potential water resources impacts from global warming in California. Presented at the International Workshop on the Impact of Global Climate Change on Energy Development in Nigeria, Lagos, Nigeria, March 28-31.

Rosen, H., and M. Reuss. 1988. The Flood Control Challenge: Past, Present, and Future. Chicago: Public Works Historical Society. 167 pp.

Sacramento Area Flood Control Agency (SAFCA). 1994a. Interim Reoperation of Folsom Dam and Reservoir. Final Environmental Impact Report/Environmental Assessment. Sacramento: SAFCA. December 15.

Sacramento Area Flood Control Agency (SAFCA). 1994b. The Lower American River Task Force, Proceedings of Phase One. Sacramento: SAFCA. July.

Sacramento Bee. September 17, 1993a. Serna Leaves No Doubts: Natomas to be developed. B1, B3.

Sacramento Bee. October 13, 1993b. Building Flood Gates Opened in Natomas. B3.

Sacramento Bee. March 24, 1994a. New fears raised about risk of capital flooding. B1, B6.

Sacramento Bee. April 20, 1994b. Bid to Give Old Fields New Life. B1.

Sacramento News and Report. January 28, 1995. Editorial: Doolittle's Dam Plan.

Sadoff, Col. L.R. 1992. Observations and Comments: WRC Environmental and M.Swanson and Assoc. Final Report... Unpublished letter to Wm. H. Edgar, Exec. Dir. SAFCA, May 6, from the District Engineer.

Sandman, P.M. 1985. Getting to maybe: Communications aspects of siting hazardous waste facilities. Seton Hall Legislative Journal 9:442-456.

Sax, J.L. 1993. Bringing an ecological perspective to natural resources law: Fulfilling the promise of the public trust. Pp. 148-161 in Natural Resources Policy and Law: Trends and Directions, L.J. MacDonnell and S.F. Bates (eds.). Washington, D.C.: Island Press.

Schumm, S.A. 1973. Geomorphic thresholds and the complex response of drainage systems. Pp. 299-310 in Fluvial Geomorphology, Morisawa, M. (ed.). Binghamton: State University of New York, Publications in Geomorphology.

Schumm, S.A. 1977. The Fluvial System. New York: John Wiley & Sons.

Schumm, S.A., M.P. Mosley, and W.E. Weaver. 1987. Experimental Fluvial Geomorphology. New York: John Wiley. 413 pp.

Schuster, R.L. and G.F. Embree. 1980. Landslides caused by rapid draining of Teton Reservoir, Idaho. Pp. 1-14 in Proceedings of the 18th Annual Engineering Geology and Soils Engineering Symposium, Boise, Idaho, April 24, 1980.

Science Applications International. 1991. Final Report of Comments on American River Watershed Investigation, WA No. C-2-56(0). For U.S. EPA, Office of External Affairs, Region 9, May 29, 1991. Golden: Science Applications International.

Secretary for Resources, California. 1973. Guidelines for Implementation of the California Environmental Quality Act of 1970.

Shabman, L. 1993. Environmental Activities in Corps of Engineers Water Resources Programs: Charting a New Direction. USACE Institute for Water Resources, IWR-93-PS-1. 85 pp.

Shabman, L., and K. Stephenson. 1992. The possibility of community-wide flood control benefits: Evidence from voting behavior in a bond referendum. Water Resources Research 28(4):959-964.

Shields, D.F. 1982 Environmental Features for Flood-Control Channels. Technical Report E-82-7. Environmental Laboratory, Waterways Experiment Station, Vicksburg, Mississippi. Prepared for Office, Chief of Engineers, USACE. Washington, D.C.:

Slovic, P. 1987. Perceptions of risk. Science 236:280-285.

Snider, W.M., and E. Gerstung. 1986. Instream flow requirements of the fish and wildlife resources of the Lower American River, Sacramento County, California. Stream Evaluation Report No. 86-1. Sacramento: California Department of Fish and Game.

St. Louis Post-Dispatch. September 26, 1993. Editorial: A Bad Deal for Taxpayers. B2.

Starr, C. 1985. Risk management, assessment and acceptability. Risk Analysis 5(2):97-102.

Stedinger, J.R. 1983a. Design events with specified flood risk. Water Resources Research 19(2): 511-522.

Stedinger, J.R. 1983b. Confidence intervals for design events. Journal of the Hydraulics Division of the ASCE 109(HY1):13-27.

Stedinger, J.R., D. Heath, and N. Nagarwalla. 1989. Event Tree Simulation Analysis for Dam Safety Problems. In Risk Analysis and Management of Natural and Man-Made Hazards, Proceedings of the Third Engineering Foundation Conference, Y.Y. Haimes and E.Z. Stakhiv (eds.). New York: American Society of Civil Engineers.

Stedinger, J.R., R.M. Vogel, and E. Foufoula-Georgiou. 1993. Frequency analysis of extreme events. Chapter 18 in Handbook of Hydrology, D.R. Maidment (ed.). New York: McGraw-Hill.

Suter, G.W., II. 1993. Ecological Risk Assessment. Chelsea, Mich.: Lewis.

Swetnam, T.W., and J.L. Betancourt. 1990. Fire-oscillation relations in the southwestern United States. Science 249:1017-1020.

Tasker, G.D., and J.R. Stedinger. 1986. Estimating generalized skew with weighted least squares regression. Journal of Water Resources Planning and Management 112(2):225-237.

Tasker, G.D., and J.R. Stedinger. 1989. An operational GLS model for hydrologic regression. Journal of Hydrology 111(1-4):361-375.

Taylor, D.W. 1937. Stability of earth slopes. Journal of the Boston Society of Civil Engineers 24(3).

Taylor, D.B., K.D. Hofseth, L.A. Shabman, and D.A. Moser. 1992. Moving toward a probability-based risk analysis of the benefits and costs of major rehabilitation projects. Pp. 148-173 in Risk-based Decision Making in Water Resources V, Y.Y. Haimes, D.A. Moser, and E.Z. Stakhiv (eds.). New York: American Society of Civil Engineers.

Taylor, B.N., and C.E. Kuyatt. 1993. Guidelines for evaluating and expressing uncertainty for NIST measurement results. NIST Tech. Note. 1297. Gaithersburg, Md.: National Institute of Standards and Technology. 15 pp.

Thomas, D.M. 1976. Flood frequency—Expected and unexpected probabilities. U.S. Geological Survey Open File Report 76-775.

Thomas, W.O. 1985. A uniform technique for flood frequency analysis. Journal of Water Resources Planning and Management 111(3):321-337.

Tung, Y.K., and L.W. Mays. 1981. Risk models for levee design. Water Resources Research 17(4).

U.S. Army Corps of Engineers (USACE). 1959. Reservoir Operation Criteria for Flood Control: Sacramento-San Joaquin Valley, Ca. Department of the Army, Sacramento District, Sacramento, Calif., October.

U.S. Army Corps of Engineers (USACE). 1982. REGFQ: Regional Frequency Computations. Hydrologic Energy Center, Davis, California, Report CPD-27, revised edition.

U.S. Army Corps of Engineers (USACE). 1987. Water Control Manual, Folsom Dam and Lake, American River, California. Department of the Army, Sacramento District, Sacramento, Ca., December.

U.S. Army Corps of Engineers (USACE). 1992a. Guidelines for Risk and Uncertainty Analysis in Water Resources Planning, Vol. I, Principles with Technical Appendices. Report 92-R-1. Fort Belvoir, Va.: Water Resources Support Center, Institute of Water Resources.

U.S. Army Corps of Engineers (USACE). 1992b. Guidelines and Procedures for Risk and Uncertainty Analysis in Corps Civil Works Planning, Vol. II: Example Cases, Guidelines for Risk and Uncertainty Analysis in Water Resources Planning, Report 92-R-2. Fort Belvoir, Va: Water Resources Support Center, Institute of Water Resources.

U.S. Army Corps of Engineers (USACE). 1992c. Streams Above the Line: Channel Morphology and Flood Control. Flood Control Channels Research Program, Proceedings, October 1992. Seattle, Wash.: USACE.

U.S. Army Corps of Engineers (USACE). 1994. Risk-based analysis for evaluation of hydrology/hydraulics and economics in flood damage reduction studies. Circular EC-1105-2-205. Washington: USACE.

U.S. Army Corps of Engineers (USACE), Sacramento District. 1991. American River Watershed Investigation, California: Feasibility Report. U.S. Army Corps of Engineers, Sacramento District, and The Reclamation Board, State of California.

U.S. Army Corps of Engineers (USACE), Sacramento District. 1993. Folsom Dam and Lake Operation Evaluation, Sacramento, Calif., December.

U.S. Army Corps of Engineers (USACE), Sacramento District. 1994a. Alternatives Report: American River Watershed. Sacramento: USACE.

U.S. Army Corps of Engineers (USACE), Sacramento District. 1994b. American River Flood Control Project Task 1: Offstream Flood Control Storage on Deer Creek, Draft Reference.

U.S. Bureau of Reclamation. 1986. Preventing a Crisis: the Operation of Folsom Dam during the 1986 Flood. Sacramento: U.S. Bureau of Reclamation.

U.S. Congress. 1966. A unified national program for managing flood losses. House doc. no. 465. 89th Cong., 2d Sess. Washington, D.C.: U.S. Government Printing Office.

U.S. Congress, House. 1994. Report of the Bipartisan Task Force on Disasters.

U.S. Department of the Interior (USDI), Bureau of Reclamation and Fish and Wildlife Service. 1993. Implementation of the Central Valley Improvement Act, Title XXXIV, Public Law 102-575. First Annual Report, September 30. Sacramento: USDI.

U.S. Department of the Interior (USDI). 1994. Draft Contract Between the United States of America and the Sacramento Area Flood Control Agency Concerning the Operation of Folsom Dam and Reservoir. Rev. NCCAO 12/13-1994. Contract No. 5-07-20-X0332.

U.S. Environmental Protection Agency (USEPA). 1992. Framework for Ecological Risk Assessment. EPA/630/R-92/001. Washington, D.C.: Environmental Protection Agency.

U.S. Environmental Protection Agency (USEPA). 1993. A Review of Ecological Assessment Case Studies from a Risk Assessment Perspective. EPA/630/R-92/005. Washington, D.C.: Environmental Protection Agency.

U.S. Environmental Protection Agency (USEPA). 1994a. A Review of Ecological Assessment Case Studies from a Risk Assessment Perspective (Vol. II). EPA/630/R-92/003. Washington, D.C.: Environmental Protection Agency.

U.S. Environmental Protection Agency (USEPA). 1994b. Water Quality Standards for Surface Waters of the Sacramento River, San Joaquin River, and San Francisco Bay and Delta of the State of California: Proposed Rule. 40 CFR Part 131. Federal Register 59(4):810-852.

U.S. Fish and Wildlife Service (USFWS). 1991. American River Watershed Investigation, Auburn Area Substantiating Report. American River Watershed Investigation Feasibility Report. In App. S, Part 1. U.S. Fish and Wildlife Service. Sacramento: USACE.

U.S. Fish and Wildlife Service (USFWS). 1994. Endangered and Threatened Wildlife and Plants; Revised Proposed Critical Habitat for the Delta Smelt, and Proposed Determination of Threatened Status for the Sacramento Splittail: Proposed rules. 50 CFR Part 17. Federal Register 59(4):852-869.

U.S. Water Resources Council (WRC), Hydrology Committee. 1967. Uniform Technique for Determining Flood Flow Frequency. Bulletin 15, December. Washington: WRC.

U.S. Water Resources Council. 1983. Economic and Environmental Principles and Guidelines for Water and Related Land Resources Implementation Studies. Washington, D.C.: U.S. Government Printing Office.

Vicens, G.J., I. Rodríguez-Iturbe, and J.C. Schaake, Jr. 1975. A Bayesian framework for the use of regional information in hydrology. Water Resources Research 11(3):405-414.

Von Thun, J.L. 1987. Use of risk-based analysis in making decisions on dam safety. In Engineering Reliability and Risk in Water Resources (NATO Advanced Study Institute, Tucson, Ariz., May 1985), L. Duckstein, and E. Plate (eds.), Dordrecht, the Netherlands: M. Nijhoff.

Water Engineering and Technology, Inc (WET). 1991. Geomorphic analysis and bank protection alternatives for Sacramento River (RM 0-78), Feather River (RM 28-61, Yuba River (RM 0-11), Bear River (RM 0-17), American River (RM 0-23), and portions of ... [others]. U.S. Army Corps of Engineers, Contract DACW05-88-D-0044, 334 pp.

Walters, C. 1986. Adaptive Management of Renewable Resources. New York: Macmillan. 374 pp.

Water Education Foundation. 1988. Layman's Guide to the American River. Sacramento: Water Education Foundation.

Water Resources Development Act of 1986. Public Law 662, 99th Cong., 2d sess. (17 November 1986).

Warner, M.L., J.L. Moore, S. Chatterjee, et al. 1974. An Assessment Methodology for the Environmental Impact of Water Resource Projects. EPA-600/5-74-016. Washington Environmental Research Center. Prepared for the Office of Research and Development. U.S. Environmental Protection Agency. Washington, D.C.

Wells, W.G., II. 1987. The effects of fire on the generation of debris flows in southern California. Pp. 105-114 in Debris Flows/Avalanches: Process, Recognition, and Mitigation, J.E. Costa and G.F. Wieczorek (eds.). Boulder, Colo.: Geological Society of America.

White, G.F. 1975. Flood Hazard in the United States: A Research Assessment. Monograph no. NSF-RA-E-75-006. Boulder, Colo.: University of Colorado Institute of Behavioral Science.

White, G.F. 1986. Strategic aspects of urban floodplain occupance. In R.W. Kates and I. Burton (eds.), Geography, Resources, and Environment. Vol. I. Selected Writings of Gilbert F. White. Chicago: University of Chicago Press.

Wieman, D.M. 1992. Letter from EPA, Region 9, Director of Office of External Affairs to D.A. Banashek, Director of Washington Level Review Center, U.S. Army Corps of Engineers. May 20.

Williams, G.P., and M.G. Wolman. 1984. Downstream effects of dams on alluvial rivers. U.S. Geological Survey Professional Paper 1286.

Williams, J.G. 1995. Report of the Special Master for Water Years 1990-1993, *Environmental Defense Fund et al. versus East Bay Municipal Utility District et al.*, Alameda County (California) Superior Court No. 425955.

Williams, P.B. 1993. Assessing the True Value of Flood Control Reservoirs: the Experience of Folsom Dam in the February 1986 Flood. Paper presented at the 1993 National Conference on Hydraulic Engineering and International Symposium on Engineering Hydrology, San Francisco, Ca., July 25-30.

Williams, P.B., and J. Galton. 1987. Analysis of the 100-Year Flood into Folsom Reservoir. Report #387, March 27, Phillip Williams & Associates, San Francisco.

Wood, E. 1978. Analyzing hydrologic uncertainty and its impact upon decision making in water resources. Advances in Water Resources 1(5):299-305.

Woolhiser, D.A., T.O. Keefer, and K.T. Redmond. 1993. Southern oscillation effects on daily precipitation in the southwestern United States. Water Resources Research 29(4):1287-1295.

WRC Environmental and M. Swanson and Associates. 1992. A review of key issues from the American River Watershed Investigation Feasibility Study and an analysis of alternatives for increasing flood protection in Sacramento. Prepared for Sacramento Area Flood Control Agency. Sacramento, CA.

Yolo Basin Foundation. 1994. Yolo Basin Wildlife Area, Time-Line of Events.

Zellner, A. 1971. An Introduction to Bayesian Inference in Econometrics. New York: John Wiley.

A
Biographical Sketches
of Committee Members

RUTHERFORD H. PLATT (Chair) is Professor of Geography and Adjunct Professor of Regional Planning at the University of Massachusetts at Amherst. He received his Ph.D. in geography in 1971 and a J.D. in law in 1967 from the University of Chicago. His B.A. is in political science from Yale University. He has served as a consultant to numerous agencies, including Federal Emergency Management Agency, U.S. Army Corps of Engineers, Office of Coastal Zone Management, Tennessee Valley Authority, and others. His NRC service includes the Committee on Flood Insurance Studies, Committee on Federal Water Research, Committee on NFIP Levee Policy, Committee on Options to Preserve Cape Hatteras Lighthouse, and Committee on Coastal Erosion Hazards.

KENNETH W. POTTER *(Vice Chair)* is Professor of Civil and Environmental Engineering and Chair of the Water Resources Management Program at the University of Wisconsin at Madison. He received his B.S. in geology from Louisiana State University in 1968 and his Ph.D. in geography and environmental engineering from The Johns Hopkins University in 1976. His teaching and research interests are in hydrology and water resources, and include estimation of hydrological risk, especially flood risk; hydrological modeling and design; stormwater modeling, management, and design; assessment of human impacts on hydrological systems; and estimation of hydrological budgets, both surface and ground water. Dr. Potter was a member of the WSTB and has participated in a number of NRC activities.

LEO M. EISEL is President, McLaughlin Water Engineers in Denver, Colorado.

He received his Ph.D. in engineering from Harvard University in 1970. From 1971 to 1973 he was a staff scientist with the Environmental Defense Fund in New York. Later, he became Director of the Illinois Division of Water Resources, and from 1977 to 1980 he was Director of the U.S. Water Resources Council. Dr. Eisel has been a member of the WSTB, the Committee to Review the Metropolitan Washington Area Water Supply Study, and the recent Committee on Western Water Management. Dr. Eisel is broadly experienced in water supply planning and hydrologic engineering.

JAMES D. HALL is Professor Emeritus in the Department of Fisheries and Wildlife, Oregon State University. He received his Ph.D. in fisheries from the University of Michigan. His research interests include population dynamics of freshwater fish, effects of watershed practices on streams, and stream ecology. He has done extensive work on anadromous fish habitat in western North America. Dr. Hall served as a visiting professor at the Institute of Animal Resource Ecology, University of British Columbia; the Department of Zoology, University of Canterbury, New Zealand; and the University of Edinburgh, Scotland.

L. ALLAN JAMES is Associate Professor at the Department of Geography, University of South Carolina. He received his Ph.D. in geography and geology and has M.S. degrees in both water resources management and geography from the University of Wisconsin-Madison and his B.A. in geography from the University of California, Berkeley. While his expertise is in the hydrogeomorphology (with specific experience studying rivers flowing out of the Sierra foothills), his research interests are interdisciplinary. His work has focused on hydraulic mining sedimentation of streams draining to the Sacramento Valley and geomorphic mapping in the northwest Sierra Nevada.

WILLIAM KIRBY holds B.C.E. (1963), M.S. (1966), and Ph.D. (1968) degrees from Cornell University in sanitary engineering, hydraulics, and applied probability. He has worked as a research hydrologist for the U.S. Geological Survey since 1967. He is now in the Office of Surface Water, where he develops and maintains procedures and computer programs for indirect discharge determinations and other hydraulic computations and develops procedures for calculating probability laws of hydrologic storage models for floods and droughts. He has had considerable experience in watershed modeling and flood-frequency analysis. Dr. Kirby's fields of specialization are random behavior and control of hydrologic systems; probabilistic structure of the streamflow process in flood and in drought; reservoir operating policies; development and evaluation of statistical procedures; one-dimensional hydraulic analysis and flow modeling; and indirect discharge determinations. He served on WSTB's Committee on Estimating the Probabilities of Extreme Floods.

NANCY Y. MOORE received her Ph.D. in water resources systems engineering with a minor in operations research and econometrics from the University of California at Los Angeles. Her research focused on the optional timing, sequencing and sizing of multiple reservoir surface water supply facilities when demand depends on price. Dr. Moore is a Senior Research Engineer, Resource Management Department, at RAND. She previously served as Director of Development and Engineer in the Engineering and Applied Sciences Department at RAND. Dr. Moore has conducted studies on efficient ground and surface water use in California, evaluating the effects of the state's water rights, institutions, pricing, and planning process on efficient use and proposed alternative ways to improve water use efficiency. She led a study of the impacts of California's 1991 drought and is leading a survey of urban water agencies on water availability and distribution during the 1987-1991 drought. Dr. Moore has written widely on water management issues, including market transfers and conjunctive use of surface and ground water. She served on NRC's Committee to Review the Glen Canyon Environmental Studies.

JOHN W. MORRIS (Lt. Gen. U.S. Army Ret.) is President, J.W. Morris Ltd. He was formerly Chief of Engineers, U.S. Army Corps of Engineers, and Chair of Construction Management at the University of Maryland. Gen. Morris also served as Executive Director for International Operations for Royal Volker Stevin N.V. and Chair/CEO of the Planning Research Corporation Engineer Group. He earned a B.S. in civil engineering from the U.S. Military Academy in 1943 and an M.S. from the University of Iowa in 1948. He is an expert in construction management and has received numerous awards and honors from professional societies and government agencies, including a Presidential Citation for Management, Construction Man of the Year (1977) from Engineering News Record, and the Pladium Medal sponsored by the Audubon Society. Gen. Morris is a member of the National Academy of Engineering, and until 1994 served on the Building Research Board, the Committee on Inspection for Quality Control on Federal Construction Projects, and the Committee on Architect-Engineer Responsibilities.

ANN L. RILEY earned her Ph.D. in environmental planning, specializing in floodplain and watershed management river restoration, hydrology, and water policy, from the University of California, Berkeley. She also holds a M.S. in landscape architecture from University of California-Berkeley. She is Executive Director of the southwest office of the Coalition to Restore Urban Waters and is active in the area of river management and restoration. Dr. Riley has extensive experience working in different aspects of government, including contract field work and research for the U.S. Geological Survey, land use planning for county governments in the Midwest, and river restoration and floodplain management for the California Department of Water Resources. She has taught courses in environmental science and floodplain management at several colleges and has been active in community organizing. She founded the Urban Creeks Council of California, a statewide environmental organization, and the National Coalition to Restore Urban Waters. Related experience includes serving as an instructor at the U.S. Army Corps of Engineers Waterways Experiment Station for workshops on the design of flood control projects.

LEONARD SHABMAN received a Ph.D. in agricultural economics in 1972 from Cornell University. He is Professor of Resource and Environmental Economics at Virginia Tech, Department of Agricultural and Applied Economics. His responsibilities include the conduct and management of a research program in resource and environmental policy analysis; classroom teaching; and undergraduate and graduate student advising. Dr. Shabman has conducted economic research over a wide range of topics in natural resource and environmental policy, with emphasis in six general areas: coastal resources management; planning, investment, and financing of water resource development; flood hazard management; federal and state water planning; water quality management, and fisheries management. He was an economic advisor to the Water Resources Council in 1977-1978 and scientific advisor to the Assistant Secretary of the Army, Civil Works, in 1984-1985. He served on the WSTB's Committee on Restoration of Aquatic Systems.

HSIEH WEN SHEN is Professor of Civil Engineering at the University of California at Berkeley. He earned his B.S. and M.S. in civil engineering from the University of Michigan, Ann Arbor, and his Ph.D. in civil engineering from the University of California, Berkeley, in 1961. Dr. Shen's major areas of research include sediment transport, water resources development, interaction between sediment movements and structures, mathematical modeling of movable bed streams, and stream ecology including developing flow control and release plans for ecological concerns. He was elected to the National Academy of Engineering for his work on the development of flow control and release plans of reservoirs to restore and enhance the ecological environments of rivers.

JERY R. STEDINGER is Professor of Civil and Environmental Engineering at Cornell University. He received his Ph.D. in engineering from Harvard in 1977, where he was a member of the Environmental Systems Program. He earned his M.S. in applied mathematics from Harvard University and his B.A. in applied mathematics from the University of California, Berkeley. Dr. Stedinger's research includes multireservoir systems analysis, risk analysis, and many topics in stochastic hydrology. He has extensive California experience conducting research for Pacific Gas & Electric Company. He is a previous winner of the NSF Presidential Young Investigator award. Dr. Stedinger served as a member of the NRC's Committee on Water Resources Research and the Committee on Safety Criteria for Dams.

Guide to Acronyms and Abbreviations

ARWI—American River Watershed Investigation
CDC—California Debris Commission
CDWR—California Department of Water Resources
CEQ—Council on Environmental Quality
CEQA—California Environmental Quality Act
EIS—Environmental Impact Statement
EBMUD—East Bay Municipal Utility District
EPA—Environmental Protection Agency
FEMA—Federal Emergency Management Agency
FWS—Fish and Wildlife Service
NFIP—National Flood Insurance Program
NAE—National Academy of Engineering
NAS—National Academy of Sciences
NEPA—National Environmental Policy Act
NRC—National Research Council
P&G—Principles & Guidelines
RM—river mile
SAFCA—Sacramento Area Flood Control Agency
USACE—U.S. Army Corps of Engineers
WSTB—Water Science and Technology Board